I0405044

The Manufacture of High Temperature Superconducting Tapes and Films

by
Kurt A. Richardson

ISBN: 1-58112-079-6

DISSERTATION.COM

1999

Copyright © 1999 Kurt A. Richardson
All rights reserved.

ISBN: 1-58112-079-6

Dissertation.com
USA • 1999

www.dissertation.com/library/112040796a.htm

UNIVERSITY OF SOUTHAMPTON

The Manufacture of High Temperature Superconducting Tapes and Films

A thesis submitted for the
Degree of Doctor of Philosophy

by

Kurt Antony Richardson

Department of Physics
Southampton

September 1996

UNIVERSITY OF SOUTHAMPTON
ABSTRACT
FACULTY OF SCIENCE
DEPARTMENT OF PHYSICS

Doctor of Philosophy
THE MANUFACTURE OF HIGH TEMPERATURE SUPERCONDUCTING TAPES AND FILMS
Kurt Antony Richardson

The potential benefits to society that might be gained if the promise of efficient energy technologies and ultra-fast computational technologies, through the development of superconductor technologies, can be realised are far reaching and exciting. However, if the opportunities arising from these exciting advances are to be grasped then difficulties in the large scale production of these brittle ceramic materials must be addressed. This thesis contributes to the body of knowledge that will enable a solution to this problem through consideration of two different manufacturing routes.

Firstly, an investigation was performed to examine the manufacture of high temperature superconducting tapes via a powder-in-tube technique. This initial study was concentrated on tapes manufactured from the recently discovered thallium-based superconducting material (Tl,Pb,Bi)-1223. Particular attention was paid to the microstructural evolution of the tapes' core with varying sinter temperature. It was found that the platelet-like as-prepared powder could be used to form superconducting tapes that displayed partial texturing - a feature not before observed in thallium-based tapes of this kind. The synthesised powder had an excellent intra-grain critical current density of 6.5×10^4 A cm^{-2} in an applied magnetic field of 1 T, at 77 K, with a superconducting transition temperature, T_C, of 110 K. Moreover, the powder was relatively simple to prepare – the achievement of reproducibility normally being a problem associated with the 'shake and bake' preparation of thallium materials. Transport critical current densities upto 5.6×10^3 A cm^{-2} in a magnetic field of 0 T, at 77 K, were measured in short lengths of tape. The problem of weak grain connectivity was observed.

The difficulties associated with the manufacture of long lengths of superconductor by the application of the powder-in-tube approach suggested that an alternative route might be more productive. A relatively new area of superconductor manufacture known as electrodeposition was therefore considered. A detailed study of the electrochemical fabrication of superconducting precursor films was performed. Fundamental measurements of the metals involved were carried out and then a range of precursor films synthesised and analysed in order to understand the electrodeposition process of multi-metal co-deposition. A range of major control parameters involved in the process where identified that affected both the intrinsic film characteristics and the process reproducibility. These parameters included ambient temperature, applied potential, deposition technique, and solution composition. A procedure was developed that yielded a highly reproducible process which allowed the production of quality films exhibiting void percentages from as low as 45 % and upto 90 %. Heat treatment of selected films resulted in high purity, highly aligned, Bi-2212 films with T_C's ~ 98 K, and multiphasic (Tl,Pb)-1223 films with T_C's ~ 115 K, and magnetic critical current densities of 4.99×10^5 A cm^{-2}, and 1.26×10^6 A cm^{-2} at 77 K, 1 T, respectively.

Acknowledgements

Though this thesis bears only the author's name many people provided valuable support in a variety of ways to ensure that I succeeded in overcoming the ordeal (!) of the past three years. Firstly, thanks must go to my supervisor, Dr. Peter de Groot for his support. I am indebted to Dr. Peter Lanchester and Prof. Phil Bartlett for their patients in reading parts of this thesis over and over again, ensuring that it made some inkling of sense - especially tenses.

I would like to also thank Maureen Freeman for all her help in organising my life from conferences to love - reminding me that being single is much more fun than getting hitched. There is a mountain of other people I should thank (and will) for their assistance and friendship over this trying period. These include: Dr Mohand Oussena (for taking an interest and giving valued encouragement), Stevie, Darko, Mitra, Rita, Tim, Greg, Jo, yet another Peter (Birkin), Emms, Georgie, Viv, Andy, another Steve, Keith, Si, and Fartum. Thanks also goes to Dave Beckett (it's amazing what goes on in the local council), Colin Miles, and Vincy for technical support and interesting anecdotes.

My mate Lushy, as well as being a good friend, has paid for my vices for the past 7 years - you must be mad - thanks anyway. Cheers to Master Slaters, and bit, for funding many jaunts to London for rather too adventurous weekends away. Even more thanks go to Caroline for love and encouragement.

Finally, and by far most importantly, total appreciation and respect has to go to *all* my parents for letting me doss at home for so long. as well as giving me a place to run away to. Thanks guys, I love you all very much.

The research presented and discussed herein was made possible with the support of an EPSRC studentship.

Contents

1. **Introduction** 1:1
   - 1.1 What is Superconductivity? - An Historical Overview 1:1
   - 1.2 Superconducting Properties 1:5
   - 1.3 Type I and Type II superconductors 1:8
   - 1.4 Theories of Superconductivity 1:9
   - 1.5 Manufacture of Superconducting Tapes and Films 1:10
     - 1.5.1 Molecular Beam Epitaxy 1:10
     - 1.5.2 Sputtering 1:11
     - 1.5.3 Thermal Spraying/Deposition 1:11
     - 1.5.4 Laser Deposition 1:11
     - 1.5.5 Sol-Gel Techniques 1:12
     - 1.5.6 Powder-in-Tube (PIT) Method 1:12
     - 1.5.7 Electrodeposition 1:13
   - 1.6 The Scope of this Work 1:13
   - 1.7 Thesis Outline 1:14
   - References 1:15

2. **Characterisation Techniques** 2:1
   - 2.1 Introduction 2:1
   - 2.2 X-ray Diffraction Spectroscopy (XRD) 2:1
     - 2.2.1 Instrumentation 2:1
     - 2.2.2 XRS Data Analysis 2:2
     - 2.2.3 Theory of Diffraction 2:3
   - 2.3 Scanning Electron Microscopy 2:3
     - 2.3.1 SEM Operation 2:4
     - 2.3.2 Sample Preparation 2:6
   - 2.4 Energy Dispersive Spectroscopy 2:7
   - 2.5 AC Susceptibility 2:8
     - 2.5.1 Experimental procedure 2:8
   - 2.6 Vibrating Sample Magnetometry 2:10
     - 2.6.1 VSM Operation 2:11
   - 2.7 Squid Magnetometer 2:11
   - 2.8 Transport Critical Current Rig 2:14
   - 2.9 Potentiostat 2:15
   - 2.10 Summary 2:17
   - References 2:18

3. **Synthesis and Characterisation of Thallium-Based Powder and Ag-Sheathed Tapes** 3:1
   - 3.1 Synthesis of the Thallium-Based Superconducting Powder 3:1
   - 3.2 The Manufacture of Thallium-Based Superconducting Tapes via the PIT Technique. 3:4
     - 3.2.1 Background 3:4
     - 3.2.2 Synthesis and Characterisation of (Tl,Pb,Bi)-1223 Powder 3:4
     - 3.2.3 Fabrication and Characterisation of (Tl,Pb,Bi)-1223 Tapes 3:7
     - 3.2.4 Results and Discussion 3:8
     - 3.2.5 Conclusions 3:17
   - 3.3 Further Work 3:17
   - References 3:18

4. **Electrochemistry and Electrodeposition of Superconductor Constituents** 4:1
   - 4.1 Introduction 4:1

4.2		Theory of Electrochemistry	4:1
	4.2.1	*Basic Electrochemistry*	*4:1*
	4.2.2	*The Double Layer*	*4:3*
	4.2.3	*Mass Transport*	*4:7*
		4.2.3.1 Diffusion	4:8
		4.2.3.2 Convection	4:8
		4.2.3.3 Migration	4:8
	4.2.4	*Concentration Profile*	*4:9*
	4.2.5	*Cyclic Voltammetry*	*4:12*
		4.2.5.1 Reversible Systems	4:14
		4.2.5.2 Irreversible and Quasi-reversible Systems	4:16
	4.2.6	*Chronoamperometry*	*4:17*
	4.2.7	*Experimental Considerations*	*4:18*
		4.2.7.1 Reference Electrode	4:18
		4.2.7.2 Solvent Selection	4:19
		4.2.7.3 Supporting Electrolyte and Complexing Agents	4:20
		4.2.7.4 Substrate Preparation	4:21
		4.2.7.5 Nitrate Salt Dehydration	4:22
		4.2.7.6 Dry Box Electrochemistry	4:23
		4.2.7.7 iR_u Drop	4:25
4.3		Electrochemistry of Superconductor Constituents	4:26
	4.3.1	*Experimental*	*4:26*
	4.3.2	*Electrochemistry of Copper*	*4:26*
	4.3.3	*Electrochemistry of Lead*	*4:31*
	4.3.4	*Electrochemistry of Thallium*	*4:34*
	4.3.5	*Electrochemistry of Bismuth*	*4:37*
	4.3.6	*Electrochemistry of Mercury*	*4:40*
	4.3.7	*Electrochemistry of Barium, Strontium, and Calcium.*	*4:43*
4.4		Electrodeposition of Superconductor Constituents	4:47
	4.4.1	*What is Electrodeposition?*	*4:47*
	4.4.2	*Experimental*	*4:48*
	4.4.3	*Electrodeposition of Cu, Pb, Tl, Bi, Ba, Sr, and Ca*	*4:48*
4.5		Conclusions	4:52
References			4:53
5	**Experimental Considerations for the Electrodeposition of Superconducting Precursor Films**		**5:1**
5.1		Introduction	5:1
5.2		Electrodeposition of High Temperature Superconductors	5:2
5.3		Experimental Considerations	5:8
	5.3.1	*Selecting the Correct Applied Potential*	*5:8*
	5.3.2	*Temperature Stability*	*5:12*
	5.3.3	*The Effect of Water*	*5:13*
	5.3.4	*Metallic Ion Depletion*	*5:14*
	5.3.5	*Substrate Selection*	*5:15*
	5.3.6	*Electrodeposition: Silent, Stirring, and Ultrasonic*	*5:16*
		5.3.6.1 Silent Deposition	5:16
		5.3.6.2 Stirred Deposition	5:17
		5.3.6.3 Deposition in the Presence of an Ultrasonic Field	5:18
5.4		Electrodeposition of Bi-Sr-Ca-Cu Films	5:23
	5.4.1	*Solution Optimisation*	*5:23*
	5.4.2	*Reproducibility and Heat Treatment*	*5:29*
5.5		Conclusions	5:31

	References	5:32
6	**Electrodeposition of Thallium-Based Superconductor Precursor Films**	**6:1**
6.1	Introduction	6:1
6.2	Experimental Procedure	6:2
6.3	Electrodeposition of Ba-Ca-Cu Films	6:3
6.4	Electrodeposition of Tl-Ba-Ca-Cu Films	6:7
6.5	Electrodeposition of Tl-Pb-Sr-Ca-Cu Films	6:9
6.5.1	Constant Potential Deposition of TPSCC Films	6:10
6.5.2	Pulsed Potential Deposition of TPSCC Films	6:12
6.5.3	Towards a Two Stage Process for the Manufacture of TPSCC films	6:15
6.6	Electrodeposition of Hg-Ba-Ca-Cu Films	6:18
6.7	Discussion and Further Work	6:19
	References	6:21
7	**Superconducting Properties of Electrodeposited Films**	**7:1**
7.1	Introduction	7:1
7.2	Superconductivity in Bi-Sr-Ca-Cu Films	7:1
7.3	Superconductivity in Tl-Pb-Sr-Ca-Cu Films	7:9
7.4	Critical Current Densities in Electrodeposited Superconducting Films	7:13
7.5	Conclusions and Discussion	7:14
	References	7:15
8	**Summary and Discussion**	**8:1**
8.1	Introduction	8:1
8.2	PIT Fabricated Tapes	8:1
8.3	Electrochemistry and Electrodeposition	8:3
8.4	Other Processing Methods	8:7
8.5	Final Comment	8:9
	References	8:10
Appendix A – The Bean Model		**A-1**
Appendix B – Thickness of an Electrodeposited Film		**B-1**
Publications List		**P-1**
Papers		P-1
Conference Presentations		P-1

Chapter 1:

Introduction

1 Introduction

1.1 What is Superconductivity? - An Historical Overview

One of the most notable events in the physical sciences this century has been the discovery of high temperature superconductors by J. George Bednorz and K. Alex Müller in April 1986 whilst working at the IBM laboratories in Ruschlikon, Switzerland [1]. The discovery was the culmination of 75 years of intensive research performed by groups around the globe whose collective aim was to raise the temperature at which a superconductor becomes superconducting to greater and greater heights. Before 1986 the highest transition temperature, T_C, achieved had been 23.2 K for Nb_3Ge in 1973 [2]. Bednorz and Müller's new compound, identified as a member of the La-Ba-Cu-O system, became superconducting at 30 K and this exciting find initiated the rush to discover further superconducting cuprate materials. This quest has yielded a variety of materials with T_C's as high as 133 K.

In 1908 a Dutch physicist named Heike Kamerlingh Onnes, of the University of Leiden, succeeded in liquefying helium for which he received the Nobel Prize for Physics in 1913. Three years later (1911) he found that, amazingly, the electrical resistance of Hg plummeted to zero at 4.2 K [3](Figure 1.1). He also noted that electric currents persisted in the Hg, in the absence of voltage, without causing the heating effects associated with resistance in conventional conductors. The physics of superconductivity was born. Subsequent searches for superconductivity, triggered by his observations, began with studies of the elements which then extended to simple alloys and intermetallic compounds.

Of all the elements studied Nb has the highest T_C of 9.3 K, but higher values have been found in intermetallic compounds. For example, in 1941, a T_C of 16.0 K was found for NbN [4]. The transition temperature gradually rose to 23.2 K in 1973 for Nb_3Ge. A number of sulphur-containing phases were also found to exhibit superconductivity, such as the $Li_xTi_{1.1}S_2$ ($x = 0.1 - 0.3$; $T_C = 10 - 15$ K) [5] and $Mo_{6-x}A_xS_6$ (A = Cu, Ag, Pb, Sn, Zn, Mg; $x = 0.9 - 1.5$; $T_C = 3 - 13$ K) [6] systems. In 1975 the first polymeric superconductor, $(SN)_x$, with $T_C < 1$ K was reported [7] and five years later, following a prediction by London forty years earlier [8], superconductivity in organic materials was discovered. Despite the prediction that a T_C of 100 K could be achieved in these materials [9] the current record stands at a lowly 8 K [10].

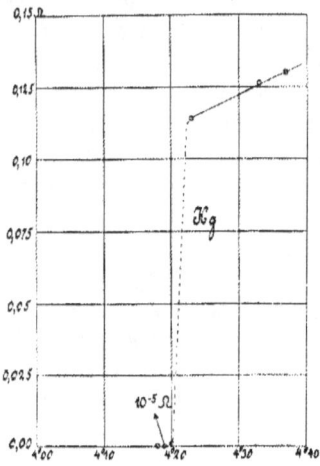

Figure 1.1 Resistance in ohms of a specimen of mercury versus absolute temperature. This plot by Kamerlingh Onnes marked the discovery of superconductivity (reference 3).

Investigations were also undertaken to observe superconductivity in metal oxides. In 1964 the first superconducting oxides were reported with T_C's ≈ 1 K; these were TiO and NbO. $SrTiO_{3-x}$ with Nb doping was found to superconduct at 0.7 K [11]. The T_C was subsequently raised to 13 K with the discovery of the spinel $LiTiO_4$ [12], and the distorted perovskite $Ba(Pb_{1-x}Bi_x)O_3$ [13]. Undoubtedly, the most important result in the field of superconductivity was the discovery of superconductivity in a copper oxide based system, specifically $La_{2-x}Ba_xCuO_4$, as mentioned above (Figure 1.2). The importance of this discovery was immediately recognised by the scientific community by awarding Bednorz and Müller the Nobel Prize for Physics, less than one year after the publication of their findings. By the end of 1986 superconductivity research had achieved revolutionary advances with the effort of Paul C. W. Chu and colleagues at the University of Houston in America. Signs of superconductivity above 77 K were repeatedly observed in the poorly-characterised samples during the period, strongly affirming the existence of superconductivity in the liquid-nitrogen temperature range. The scientific world knew that the textbooks had to be rewritten after January 1987, when the Houston group, in collaboration with M. K. Wu, then at the University of Alabama at Huntsville, achieved stable and reproducible superconductivity in $YBa_2Cu_3O_{7-\delta}$ (Y123) (Figure 1.3), with a T_C close to 100 K [14]. Superconductivity at such high temperatures defied the common understanding of solids and as of 1996 the theory of high temperature superconductivity has yet to be formulated.

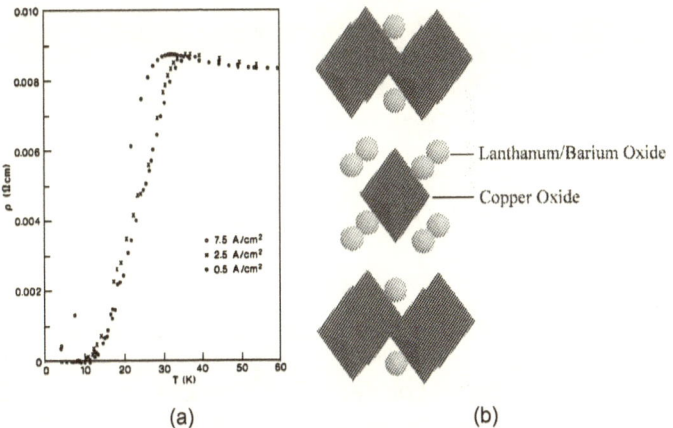

Figure 1.2 The first evidence of high temperature superconductivity. (a) The resistance versus temperature for $(Ln,Ba)_2CuO_4$, and (b) the crystal structure of $(Ln,Ba)_2CuO_4$.

Until 1987 the inconvenience, and expense, of liquid helium refrigeration meant that the widespread application of wires and tapes was considered economically unfeasible. However, with the advent of Y123 there were significant savings resulting from the displacement of liquid helium by liquid nitrogen for cooling. The race for new superconductors with higher T_C's continues. Bismuth- and thallium-based superconductors were discovered in 1988 [15,16] which become superconducting at 110 K and 125 K, respectively. The mercury-based compounds were discovered in 1993 [17], with transition temperatures upto 164 K under a pressure of 10^5 Pa. Many laboratories throughout the world have reported evidence for superconductivity at much higher temperatures, but these however, have yet to be confirmed conclusively.

Chapter 1: Introduction

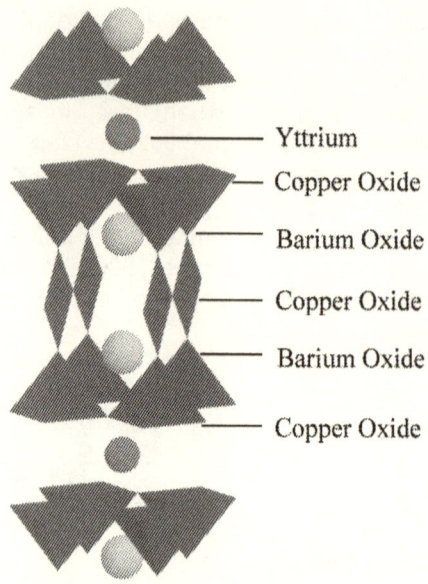

Figure 1.3 *The crystal structure of Y123.*

Very recently, three other classes of superconductor have been discovered, namely: fullerides, borate-/fluorine-containing compounds, and oxycarbonates. The fullerides, such as MC_{60}, are a totally unique family based on the curious carbon molecule Buckminster Fullerene, having T_C's of 18 K, 28 K, and 32.5 K for M = K_3, Rb_3, and Cs_2Rb, respectively [18]. Boron- and fluorine-containing compounds such as $LnSr_2Cu_{3-x}BO_7$ [19] and $Sr_{2-x}A_xCuO_2F_{2-\delta}$ (A = Ca or Ba) [20], members of the layered cuprate family, have T_C's of upto 64 K. Finally, the oxycarbonates, for example $(Y,Ca)_n(Ba,Sr)_{2n}Cu_{3n-1}O_{7n-3}CO_3$ (n = 2 - 4) [21], show yet another variation on the layered cuprate system.

Even though the liquid nitrogen barrier has been broken, many of the great promises of superconductivity technology have yet to be realised. The difficulties with the materials can be attributed to many of the material and engineering problems of high temperature superconductors (HTSC's), e.g. making long HTSC wire that can carry large currents without energy loss and can retain excellent properties over long periods of time without chemical and physical degradation. After the discovery of the transistor in 1947, it took almost 40 years to introduce the one megabit memory chip which is vital to modern computers. Modern discoveries in superconductivity go far beyond piecemeal improvements in electric devices. Future applications that take advantage of the fascinating phenomena of superconductivity will cause significant changes in electricity generation, data transmission and storage; impacts in microelectronics, communication and computers; and advances in solid state science. This progress will only be achievable through determination and persistence, and this thesis is a contribution to that endeavour.

1.2 Superconducting Properties

We can usually associate the remarkable properties of superconductors with three words: zero, infinite and perfect. The most striking characteristic of a superconductor is the abrupt disappearance of electrical resistance below a critical temperature T_C - the state of infinite electrical conductance. Below T_C the material is said to be in its superconducting state and at a temperature above T_C the material is in its normal state. The absence of electrical resistance, however, does not completely define the superconducting state. In 1933 Walter Meissner and Reiden Ochsenfeld of the University of Leiden discovered that superconductors are perfectly diamagnetic as well as being perfect conductors [22]. This means that an applied field is completely excluded from the interior of a superconductor up to a critical field, H_C, whereupon the material resorts to its 'normal' metallic state. This differs from the description of a 'perfect' conductor. The behaviour of a perfect conductor (Figure 1.4), when subjected to a magnetic field at different temperatures, depends on the state of the conductor when the field is applied. If a perfect conductor, initially in zero magnetic field (a), is cooled below T_C (b) and moved into a magnetic field (c), then by Faraday's law of induction, eddy or surface currents are set up so as to repel the flux from the interior of the material. If however a magnetic field were applied before the temperature was reduced below T_C (e), the flux would still penetrate the whole of the conductor (f). After the applied field has been removed currents will flow in the surface of the conductor so that the magnetic flux through the sample will not change (g).

Chapter 1: Introduction

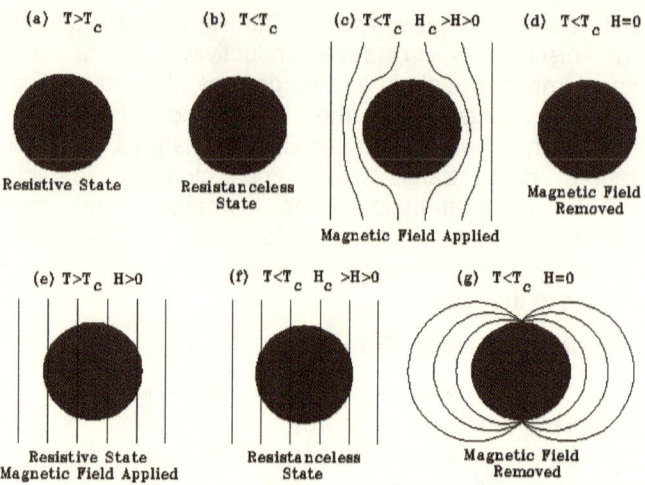

Figure 1.4 *The magnetic behaviour of a 'perfect' conductor.*

Meissner and Ochsenfeld found that when a superconductor is cooled below T_C in a magnetic field the flux is spontaneously expelled because surface currents flow that repel the field inside (Figure 1.5). The same effect is obtained if a magnetic field is applied to a superconductor below T_C. Removal of the applied field causes the persistent surface currents to disappear and hence the field associated with them (c) unlike the case for a perfect conductor. It follows from the Meissner effect that if the field inside the superconductor is zero the magnetic susceptibility, χ, of a superconductor (M/H) must be exactly -1. This is known as perfect diamagnetism.

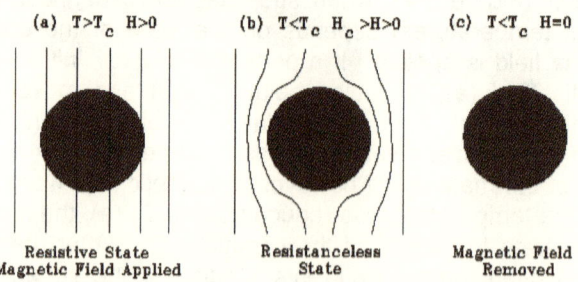

Figure 1.5 *The magnetic behaviour of a superconductor.*

The critical field, H_C, mentioned above, is observed to be related to the critical temperature T_C by the expression:

Chapter 1: Introduction

$$H_C = H_0\left[1-\left(\frac{T}{T_C}\right)^2\right] \quad \{1.1\}$$

where H_0 is the value of the critical field at 0 K (Figure 1.6). If the applied magnetic field is derived from an electrical current flowing in the superconductor, there is a corresponding critical current density, J_C, such that if $J > J_C$ then the material goes normal.

The abrupt change in resistivity at T_C in superconductors is only one of many abrupt changes that occur in the material. Figures 1.7a and b depict the temperature dependence of various selected properties of a superconducting material. Note that all the significant changes occur at T_C, and all highlight the striking changes that occur when a superconducting material passes from its normal state to its superconducting state.

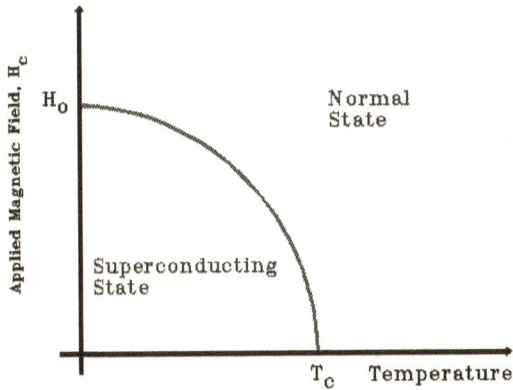

Figure 1.6 Typical relationship observed between the critical field and T_C for the elements.

Probably the most well-known superconducting property is the fact that the DC resistivity disappears at the critical temperature and remains zero below T_C. This is very different from conventional metals. In ordinary metals the resistivity, ρ, is proportional to T^5 at low temperatures. Also there is a remnant resistance as temperature approaches zero, i.e. the resistance never completely disappears.

Chapter 1: Introduction

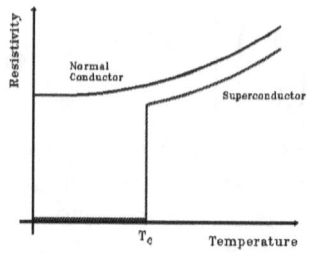
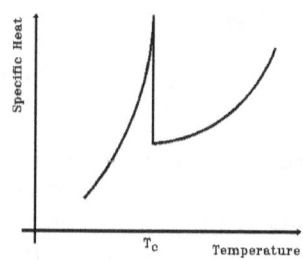

Figure 1.7a The temperature dependence of resistivity and specific heat in superconductors.

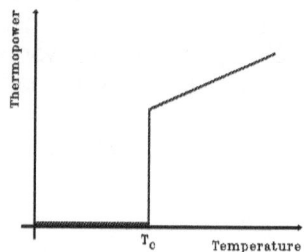
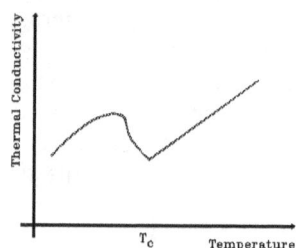

Figure 1.7b The temperature variation of thermopower and thermal conductivity in superconductors.

1.3 Type I and Type II superconductors

The magnetisation curve expected for a superconductor is shown in Figure 1.8. Pure specimens of many materials exhibit this behaviour; they are referred to as type I superconductors or, formerly, soft superconductors. The values of H_C are always too low for type I superconductors to have any useful technical application, in coils for superconducting magnets for example.

The second type of superconductors exhibit magnetisation curves of the type depicted in Figure 1.9. These are known as type II superconductors. Materials of this type tend to be alloys or transition metals with high values of electrical resistance in the normal state. Type II superconductors have superconducting electrical properties up to a field, H_{C2}[‡]. However, between the lower critical field, H_{C1}, and the upper critical field, H_{C2}, the Meissner effect is incomplete (i.e. $B_{internal} \neq 0$) and magnetic flux penetrates the superconductor. In this region the

---

[‡] H_{C2} may be 100 times higher than the value of the critical field, H_C, calculated from the thermodynamics of the transition.

1:8

superconductor is said to be in a vortex state, also known as the mixed state. An upper critical field of 41 T has been attained in an alloy of Nb, Al, and Ge at the boiling point of helium. It should be noted that in figure 1.9 a critical field H_C has been labelled. This critical field is the field as determined from thermodynamic considerations.

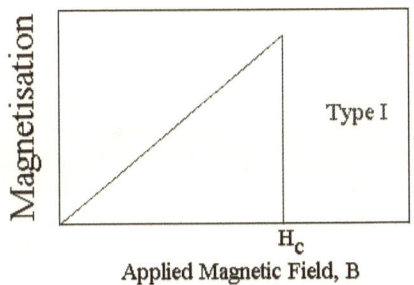

Figure 1.8 *Magnetisation versus applied magnetic field for a bulk superconductor exhibiting a complete Meissner effect (perfect diamagnetism): a type I superconductor.*

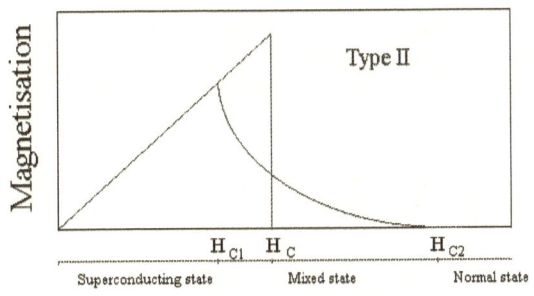

Figure 1.9 *Magnetisation versus applied magnetic field for a bulk superconductor exhibiting an incomplete Meissner effect: a type II superconductor.*

1.4 Theories of Superconductivity

A microscopic theory for conventional superconductivity was proposed in 1957 by Bardeen, Cooper, and Schrieffer [23,24] and is now known as BCS theory. The theory followed studies by Cooper of the "Cooper pair" where it was found that under certain conditions the ground state of energy of a pair of electrons is lower than that of the two free electrons. The central result of the BCS theory is this energy gap of which the critical temperature, the thermal properties, and the magnetic properties are a consequence. The weak coupling of electrons is achieved by the mechanism of Frohlich interactions. Here thermal lattice vibrations (phonons) interact with the electrons causing a weak attractive force between a pair (the typical separation distance of a pair of electrons is of the order of 10^4 Å).

The theory accounts well for the equilibrium properties of conventional or low T_C superconductors and thus values for T_C, H_C, the specific heat, and the penetration depth can be accurately derived. Since the discovery of the new high-T_C superconducting oxides many new theories have been proposed to explain the high T_C's and observed physical properties. These include several theoretical non-phononic models, such as Anderson's resonating valence bond (RVB) model [25], the spin bag model by Schrieffer et al. [26], the charge excitation model (Varma et al. [27]), plasmon exchange models [28-30], and Wilson's direct on-site electron-pair coupling mechanism [31].

1.5 Manufacture of Superconducting Tapes and Films

One of the most important present aims in the field of superconductivity is the ability to produce continuous lengths of superconductor for applications ranging from current carriers, to superconducting high-field magnets. A variety of techniques have been employed in preparing superconducting tapes and films with varying degrees of success. These include molecular beam epitaxy [32], thermal deposition of metal carboxylates (dip coating) [33], spray-drying [34], ion beam sputtering [35], RF magnetron sputtering [36], sol-gel processing [37], laser ablation [38], and metalorganic deposition [39], powder-in-tube (PIT), and electrodeposition.

1.5.1 Molecular Beam Epitaxy

Electron beam evaporation was the first technique used in depositing HTSC films [40-43]. Separate sources for each of the metal components are required. Usually, e-beam guns with independent rate control and feedback of the sources are used. Technical difficulties arise, though, because of the need for high oxygen pressures which affects the functioning of the e-beam guns, and the mass spectrometers as well as the microbalances used for process control. Techniques designed to overcome these difficulties introduce such levels of complexity that few cases are known to work successfully.

Molecular beam epitaxy (MBE) has so far proved to be one of the most complicated, expensive, and least successful techniques. The reason for pursuing MBE work on these complicated oxide systems is the possibility of acquiring greater understanding of the growth mechanism for the production of high quality crystals, but it is unlikely that this is a candidate for large scale fabrication of superconducting films.

1.5.2 Sputtering

Sputter deposition - either from diode sources, conventional magnetrons, ion-beam sources, and novel cathodes - has been widely used for making HTSC's [44-59]. Films can be sputtered either from multiple sources (elements, alloys, oxides) or from single sources. Sputter deposition offers an easy rate control. If non-reactive gases are used, the deposition rate of a sputter gun is almost proportional to the power of the electrical discharge and no feedback loop is needed to control the gun. However, the technique offers very low growth rates ~ 1.0 nm min^{-1}, and the extension to a continuous process is not clearly apparent.

1.5.3 Thermal Spraying/Deposition

This technique yields high deposition rates, ~ 1.0 μm min^{-1}, and has possible applications for a continuous deposition process, as well as for large areas. Deposited films are not epitaxial but can be highly textured achieving good critical current densities, e.g. 6×10^5 A cm^{-2} at 77 K for Y123 [33].

1.5.4 Laser Deposition

Pulsed laser deposition (PLD) is a relatively new technique [60-63]. As in the one-target sputtering process, a bulk sample forms the source. A ultra violet or visible high-energy laser pulse of 10^8 - 10^{10} W cm^{-2} and a duration of the order of 10 ns vaporises the material. A plume of constituents is ejected from the target and impinges on a substrate. The wavelength and the laser fluence have a decisive influence on the quality of the deposits. The highest quality Y123 epitaxial films have been grown to date using this technique. Problems do occur in depositing thallium-based films or lead-doped films because of their high volatility. The approach is, however, scalable.

1.5.5 Sol-Gel Techniques

Sol-gel techniques [64,65] and co-precipitation methods starting from solutions of the respective cations have been studied to avoid nitrate and carbonate contamination of grain boundaries, which is common in vapour deposition techniques using carbonate and nitrate targets. There is good control over the film content and cation distribution. High purity ceramics have been manufactured this way.

1.5.6 Powder-in-Tube (PIT) Method

PIT is a method whereby fully-reacted or partially-reacted powders of superconducting material are packed into Ag or Ag alloy tubes. The tubes are then drawn out to form wires, and then rolled to obtain tapes. After extensive research in short lengths of superconducting tape, tapes ~ 1 km in length and longer have been produced. The key problem with this approach is that homogeneity along the tape/wire is difficult to achieve. The root of this problem is the difficulty in preparing 100 % pure starting powders. If a normal state 'barrier' occurs along the tape then current carrying properties of the conductor are greatly affected. This has been overcome by producing multi-filamentary cores giving the super-current alternative routes. Texturing (grain alignment) is also hard to achieve, except in $Bi_2Sr_2Ca_2Cu_3O_{10+\delta}$ (Bi-2223) based tapes due to the favourable morphology of this material. A great deal of work into this technique [66,83] has yielded promising results from mainly Y-123, Bi-2223, Bi-2212, and Tl-2223 based systems. Y-123 tape with a critical current density, J_C, of 3.33×10^3 A cm^{-2} at 77 K and in zero field has been fabricated [67]. The samples displayed very poor in-field behaviour though - the J_C fell to 1/100th of its zero field value with the application of only 20 mT. Bi-2212 wires have been prepared [68] with J_C's upto 1.2×10^3 A cm^{-2} in zero field at 77 K.

Wires prepared from the Bi-2223 material have yielded the most promising results thus far, with J_C's over 5.4×10^4 A cm^{-2} at 77 K in zero field with impressive J_C's of ~ 1.2×10^4 A cm^{-2} in an applied field of 1 T [77]. Despite the engineering problems with this method good results have been obtained and the area remains a very active research environment. A major application of Bi-2223 tapes is the manufacture of very high field superconducting magnets at 4.2 K.

Chapter 1: Introduction

1.5.7   Electrodeposition

Electrodeposition involves the deposition of relevant metallic cations, from a deposition solution, onto a conductive substrate by electrolysis [84]. The principle attraction of the technique is its already proven scalability. High quality, thick, films may also be obtained. This method is fully reviewed in Chapter 5.

One of the important properties of a superconducting film is it's transport critical current density. Films and tapes with extremely large critical current densities have already been prepared. However, these films are normally very thin (< 1 μm) and therefore the critical currents are small. If high critical current films are to be manufactured then methods that can synthesise much thicker films must be developed. The ability to form thick films must also be accompanied by the ability to fabricate continuous lengths of conductor.

1.6   The Scope of this Work

This thesis is concerned with the manufacture of HTSC films and tapes, in particular via the PIT and electrochemical approaches. Initially, the application of the PIT method to the fabrication of tapes from a new thallium-based superconductor, based on the Tl-1223 system, was explored. The single layer Tl-1223 system was selected because of it's low anisotropy and the low dependence of transport properties upon applied magnetic fields - both positive attributes when considering tape and film manufacture. Particular attention was paid to the morphological evolution of the tapes' cores. The problem of poor grain connectivity was identified and solutions proposed. The difficulties posed by this approach when considering a continuous process, such as tape homogeneity, steered the research toward a more versatile technique - electrodeposition.

The main subject in this work is the electrodeposition of superconducting films with particular emphasis on the optimisation of the deposition process, as opposed to the optimisation of the post-deposition heat treatment. This is not a well researched area (as can be seen by the review in Chapter 5), and hence the research herein deals with the mechanisms of the deposition process in order to achieve a high standard of reproducibility in producing a variety of superconductor precursor films. The heat-treatment of the as-deposited films has been determined, and results are presented for the characteristics of the resulting superconducting films. Tl-1223 materials are difficult to make, when compared to many other systems, because of the volatility of thallium. For this reason, electrodeposited Bi-2212 samples were prepared initially in order to demonstrate the deposition process. Toward the conclusion of the research attempts were made to fabricate, for the first time via electrodeposition, (Tl,Pb)-1223 films. Superconductivity was successfully observed in both the above types of film, though no critical current densities were successfully determined.

1.7 Thesis Outline

The thesis proceeds with a description of the experimental equipment used to manufacture and characterise the samples fabricated. Chapter 3 reviews the literature for progress in PIT tape production and then describes the manufacture and characterisation of a new type of thallium-based tapes. The chapter closes with suggestions concerning future development.

Chapters 4, 5, and 6 all deal with the electrodeposition of superconductor precursor films. This starts with an investigation into the basic electrochemistry of the superconductor components, and then progresses onto the electrochemical synthesis of Bi-Sr-Ca-Cu, Tl-Ba-Ca-Cu, Ba-Ca-Cu, and Tl-Pb-Sr-Ca-Cu, and Hg-Ba-Ca-Cu superconductor precursor films. Experimental considerations were extensively researched with the implications for mass production kept in mind.

The penultimate chapter (7) deals with the superconducting properties of the heat treated films. Properties analysed were x-ray spectra (which also yielded data indicating the alignment and purity of the fabricated films), film morphology, superconducting transition temperature, and finally, the magnetic current carrying properties.

Finally, chapter 8 summarises the principal results obtained during the research, and an indication to the future of continuous superconductor production is presented.

References

1. J. G. Bednorz and K. A. Müller, *Z. Phys. B*, 64 (1986) 189.
2. J. R. Gavaler, *Appl. Phys. Lett.*, 23 (1973) 480.
3. H. Kamerlingh Onnes, *Akad. van Wetenschappen (Amsterdam)*, 14 (1911) 113.
4. G. Ascherman, E. Friederich, E. Justi, and J. Kramer, *Phys. Zeir*, 42 (1941) 349.
5. H. E. Barz, A. S. Copper, E. Corenzwit, M. Marezio, B. T. Matthias, and P. H. Schmit, *Science*, 175 (1972) 884.
6. R. Cheverel, M. Sergent, and J. Prigent, *J. Solid State Chem.*, 3 (1971) 515.
7. R. L. Greene, P. M. Grant, and G. B. Street, *Phys. Rev. Lett.*, 34 (1975) 89.
8. F. J. London, *J. Chem. Phys.*, 5 (1937) 837.
9. A. R. West, "Solid State Chemistry and Its Applications", pp. 687-9, Wiley, 1984.
10. V. N. Laukhin, E. E. Kostyuchenko, Y. V. Sushko, I. F. Shchegolev, and E. B. Yagubskii, *JETP Lett.*, 41 (1985) 81.
11. J. J. Schooley, W. R. Hosler, and M. L. Cohen, *Phys. Rev. Lett.*, 12 (1964) 474.
12. D. C. Johnson, H. Prakash, W. H. Zachariasen, and R. Viswanathan, *Mat. Res. Bull.*, 8 (1973) 777.
13. A. W. Sleight, J. L. Gillson, and P. E. Bierstedt, *Solid State Comms.*, 17 (1975) 27.
14. M. K. Wu, J. R. Ashburn, C. J. Torng, P. H. Hor, R. L. Meng, L. Gao, Z. J. Huang, Y. Q. Wang, and C. W. Chu, *Phys. Rev. Lett.*, 58 (1987) 908.
15. C. Michel, M. Hervieu, M. M. Borel, A. Grandin, F. Deslandes, J. Provost, and B. Raveau, *Z. Phys.*, B68 (1987) 421.
16. Z. Z. Shen and A. M. Hermann, *Nature*, 332 (1988) 138.
17. S. N. Putilin, E. V. Antipov, O. Chmaissem, and M. Marezio, *Nature*, 362 (1993) 226.
18. M. J. Rosseinsky and D. W. Murphy, *Chemistry in Britain*, 30 (1994) 746.
19. J. P. Chapman and J. P. Attfield, *Physica C*, 235 (1994) 351.

20. P. R. Slater, J. P. Hodges, M. G. Francesconi, P. P. Edwards, C. Greaves, I. Gameson, and M. Slaski, *Physica C*, 253 (1995) 16.
21. B. Raveau, M. Huve, A. Maignan, M. Hervieu, C. Michel, B. Domenges, and C. Martin, *Physica C*, 209 (1993) 163.
22. W. Meissner and R. Ochsenfeld, *Naturwissenschaffen*, 21 (1933) 787.
23. J. Bardeen, L. N. Cooper, and J. R. Schrieffer, *Phys. Rev.*, 106 (1957) 162.
24. J. Bardeen, L. N. Cooper, and J. R. Schrieffer, *Phys. Rev.*, 106 (1957) 1175.
25. P. W. Anderson, *Science*, 235 (1987) 1196.
26. J. R. Schrieffer, X.-G. Wen, and S. C. Zhang, *Phys. Rev. Lett.*, 60 (1988) 944.
27. C. M. Varma, S. Schmitt-Rink, and E. Abrahams, *Solid State Comms.*, 62 (1987) 681.
28. V. Z. Kresin, *Phys. Lett.*, 122A (1987) 434.
29. V. Z. Kresin and H. Morawitz, *Phys. Rev. B*, 37 (1988) 7854.
30. J. Ashkenazi, C. G. Kuper, and R. Tyk, *Solid State Comms.*, 63 (1987) 1145.
31. J. A. Wilson, *J. Phys. C*, 21 (1988) 2067.
32. R. Cabanal, J. P. Hirtz, C. Giovannelia, D. Dubreuil, G. Creuzet, J. Seijka, and G. Vizkeletty, *Processing and Applications of high T_C Superconductors*, May 1988, Proceedings of the Metallurgical Society, Warrendale, PA, 1988, pp. 3-8.
33. M. Klee and J. W. C. de Vries, *Processing and Applications of high T_C Superconductors*, May 1988, Proceedings of the Metallurgical Society, Warrendale, PA, 1988, pp. 9-15.
34. N. Nakamura, T. Nakano, S. Gotoh, and M. Shimotomai, *Processing and Applications of high T_C Superconductors*, May 1988, Proceedings of the Metallurgical Society, Warrendale, PA, 1988, pp. 17-22.
35. A. F. Hebard, R. H. Eick, A. T. Fiory, A. E. White, and K. T. Short, *Processing and Applications of high T_C Superconductors*, May 1988, Proceedings of the Metallurgical Society, Warrendale, PA, 1988, pp. 23-30.

36. P. H. Ballentine, A. M. Kadin, J. Argana, R. C. Rath, and P. McClusky, *Processing and Applications of high T_C Superconductors*, May 1988, Proceedings of the Metallurgical Society, Warrendale, PA, 1988, pp. 31-42.
37. M. Nagano and M. Greenblatt, *Processing and Applications of high T_C Superconductors*, May 1988, Proceedings of the Metallurgical Society, Warrendale, PA, 1988, pp. 43-54.
38. R. A. Neifield, S. Gunapala, G. Liang, S. A. Shaheen, M. Croft, J. Price, D. Simons, W. T. Hill, III, D. Ginley, R. Pfeffer, W. Savin, and C. Wrenn, *Processing and Applications of high T_C Superconductors*, May 1988, Proceedings of the Metallurgical Society, Warrendale, PA, 1988, pp. 61-72.
39. J. V. Mantese, A. L. Micheli, A. H. Hamdi, and R. W. Vest, *J. Mater. Sci. Res.*, (1989) 48.
40. P. Chaudari, R. H. Kock, R. B. Laibowitz, T. R. McGuire, and R. J. Gambino, *Phys. Rev. Lett.*, 58 (1987) 2687.
41. M. Naito, R. H. Hommond, B. Oh, M. R. Hahn, J. W. P. Hsu, P. Rosenthal, A. F. Marshall, M. R. Beasley, T. H. Gaballe, and A. Kapitulnik, *J. Mater. Res.*, 2 (1987) 713.
42. P. Berberich, J. Tate, W. Dietsche, and H. Kinder, *Appl. Phys. Lett.*, 53 (1988) 925.
43. R. B. Laibowitz, R. H. Koch, P. Chaudari, and R. J. Gambino, *Phys. Rev.*, B35 (1987) 8821.
44. M. Hong, S. H. Liu, J. Kwo, and B. A. Davidson, *Appl. Phys. Lett.*, 51 (1987) 694.
45. Y. Enemoto, T. Murakami, M. Suzuki, and K. Moriwaki, *Jpn. J. Appl. Phys.*, 26 (1987) L1248.
46. R. L. Sandstorm, W. J. Gallagher, T. R. Dinger, R. B. Laibowitz, and R. J. Gambino, *Appl. Phys. Lett.*, 53 (1988) 444.
47. J, Schubert, U. Popper, and W. Sybertz, *J. Less-Common Met.*, 151 (1989) 277.
48. H. C. Li, G. Linker, F. Ratzel, R. Smithey, and J. Geerk, *Appl. Phys. Lett.*, 52 (1988) 1098.
49. H. Itozaki, K. Hikagi, K. Harada, S. Tanaka, S. Yazu, and K. Tada, *Physica C*, 162-164 (1989) 367.
50. K. Kuroda, K. Kojima, M. Tanoiku, K. Yokoyama, and Hamanaka, *Jpn. J. Appl. Phys.*, 28 (1989) 1586.

51. H. Raffy, J. Arabski, A. Vaurès, S. Megtert, R. Reich, and P. Monod, *J. Less-Common Met.*, 151 (1989) 385.
52. M. Takano, J. Takada, K. Oda, H. Kitaguchi, Y. Miura, Y. Ikeda, Y. Tomii, and H. Mazaki, *Jpn. J. Appl. Phys.*, 27 (1988) 1041.
53. S. Labdi, H. Raffy, S. Megtert, A. Vaurès, and P. Tremplay, *J. Less-Common Met.*, 164/165 (1990) 687.
54. J. Geerk, X. X. Xi, Q. Li, H. C. Li, R. -L. Wang, G. Linker, O. Meyer, F. Ratzel, R. Smithey, B. Obst, and H. Reiner, *High-Temperature Superconductors, Materials Aspects* (Proc. of the ICMC '90) (Edited by H. C. Freyhardt, R. Flukiger, and M. Peuckert), Vol. 1, p 3. DGM Informationsgesellschaft, Oberusel (1991).
55. G. Baumgardt, A. Hamerich, B. Meyer, J. Müller, and R. Wunderlich, VDI-TZ Proceedings *Supraleitung und Tieftemperaturtechnik*, p. 257. VDI, Düsseldorf (1991).
56. B. David, O. Dössel, J. -P. Krummer, and M. Knechtel, VDI-TZ Proceedings *Supraleitung und Tieftemperaturtechnik*, p. 265. VDI, Düsseldorf (1991).
57. H. Downar, U. Kaufmann, T. Peterreins, and H. J. Stadler, VDI-TZ Proceedings *Supraleitung und Tieftemperaturtechnik*, p. 274. VDI, Düsseldorf (1991).
58. T. Scherer. R. Herwig, P. Marienhoff, M. Neuhaus, A. Vogt, and W. Jutzi, VDI-TZ Proceedings *Supraleitung und Tieftemperaturtechnik*, p. 319. VDI, Düsseldorf (1991).
59. H. Schmidt, O. Eibl, and B. Jobst, , *High-Temperature Superconductors, Materials Aspects* (Proc. of the ICMC '90)(Edited by H. C. Freyhardt, R. Flükiger, and M. Peuckert), Vol.1, p 285. DGM Informationsgesellschaft, Oberusel (1991).
60. D. Dijkamp, T. X. D. Wu, S. A. Haheen, N. Jisrawi, Y. H. Min Lee, W. L. McLean, and M. Croft, *Appl. Phys. Lett.*, 51 (1987) 619.
61. B. Roas, L. Schultz, and G. Endres, *Appl. Phys. Lett.*, 53 (1988) 1557.
62. G. Koren, A. Gupta, E. A. Giess, A. Segmüller, and R. B. Laibowitz, *Appl. Phys. Lett.*, 54 (1989) 1054.
63. J. Frohlingsdorf, W. Zander, and B. Stritzker, *Solid State Commun.*, 67 (1988) 965.

64. T. Monde, H. Kozuka, S. Sakka, *Chem. Lett.* 287 (1988).
65. G. Moore, S. Kramer, and G. Kordas, *Mater. Lett.*, 7 (1989) 418.
66. Y. Yamada, N. Fukushima, S. Nakayama, H. Yoshino, and S. Murase, *Jap. J. Appl. Phys.*, 26 (1987) L865.
67. M. Okada, A. Okayma, T. Morimoto, T. Matsumoto, K. Aihara, and S. Matsuda, *Jap. J. Appl. Phys.*, 27 (1988) L185.
68. M. Okada, R. Nishiwaki, T. Kamo, T. Matsumoto, K. Aihara, S. Matsuda, and M. Seido, *Jap. J. Appl. Phys.*, 27 (1988) L2345.
69. K. Heine, J. Tenbrink, and M. Thoner, *Appl. Phys. Lett.*, 55 (1989) 2441.
70. T. Asano, Y. Tanaka, M. Fukutomi, and H. Maeda, *Jap. J. Appl. Phys.*, 29 (1990) L1066.
71. K. Sato, T. Hikata, and Y. Iwasa, *Appl. Phys. Lett.*, 57 (1990) 1928.
72. Z. -W. Qi, Q. -Y. Peng, X. -Y. Long, Y. -S. Wu, M. -Y. Yi, H. Zhang, J. -P. Chen, and G. W. Yang, *Cryogenics*, 30 (1990) 855.
73. M. Okada, T. Yuasa, T. Matsumoto, K. Aihara, M. Seido, and S. Matsuda, *Jap. J. Appl. Phys.*, 29 (1990) 2732.
74. K. -I. Sato, N. Shibuta, H. Mukai, T. Hikata, M. Ueyama, and T. Kato, *J. Appl. Phys.*, 70 (1991) 6484.
75. K. -I. Sato, N. Shibuta, H. Mukai, T. Hikata, M. Ueyama, and T. Kato, *Physica C*, 190 (1991) 50.
76. M. Okada, T. Nabatame, T. Yuasa, K. Aihara, M. Seido, and S. Matsuda, *Jap. J. Appl. Phys.*, 30 (1991) 2747.
77. M. Ueyama, T. Hikata, T. Kato, and K. -I. Sato, *Jap. J. Appl. Phys.*, 80 (1991) L1384.
78. P. Haldar, J. G. Hoehn, Jr., J. A. Rice, and L. R. Motowidlo, *Appl. Phys. Lett.*, 60 (1992) 495.
79. J. -I. Shimoyama, N. Tomita, T. Morimoto, H. Kitaguchi, H. Kumakura, K. Togano, H. Maeda, K. Nomura, and M. Seido, *Jap. J. Appl. Phys.*, 31 (1992) L1328.
80. J. C. Bowker and G. A. Whitlow, *Jap. J. Appl. Phys.*, 32 (1993) 51.
81. M. Lelental, T. N. Blanton, C. L. Barnes, and R. C. Bowen, *Physica C*, 217 (1993) 79.

82. M. Yoshida and A. Endo, *Jap. J. Appl. Phys.*, 32 (1993) L1509.
83. G. Grasso, A. Perin, B. Hensel, and R. Flukiger, *Physica C*, 217 (1993) 335.
84. R. N. Bhattacharya and R. D. Blaugher, *Thallium-Based High-Temperature Superconductors* (Edited by A. M. Hermann, and J. V. Yakhmi), pp. 279-287, Marcel Dekker, New York, 1994.

Chapter 2:
Characterisation Techniques

2 Characterisation Techniques

2.1 Introduction

A variety of techniques have been used to characterise the prepared samples to determine composition, morphology, crystal structure, and superconducting properties. In this chapter the techniques, and equipment, used to obtain such data are described. For the characterisation of the superconducting PIT tapes x-ray spectroscopy, scanning electron microscopy, AC susceptibility, transport critical current rig, and vibrating sample magnetometry were employed to determine structure, superconducting transition temperature, intrinsic current carrying properties, and transport current carrying properties. Similarly, for the analysis of the electrodeposited films, the same techniques were used (except the critical current rig) in conjunction with energy dispersive spectroscopy and SQUID magnetometry. The films were manufactured using a EG&G Model 273A potentiostat / galvanostat.

2.2 X-ray Diffraction Spectroscopy (XRD)

X-ray diffraction is one of the most useful techniques available in solid-state physics for the characterisation of crystalline materials. XRD may be employed in a variety of applications yielding detailed information on crystal structure, particle size, phase transitions and to a lesser extent, crystal defects and disorder. The use of XRD in this work was in the identification of the superconducting phase formed. For the electrodeposited films no special preparation was required to perform x-ray diffractometry. The superconducting powder used for the preparation of the PIT samples were polycrystalline.

2.2.1 Instrumentation

The Phillips system is a conventional θ-2θ diffractometer shown in Figure 2.1. An x-ray tube provides copper - K_α radiation that is collimated through an aperture diaphragm on to the sample. The diffracted radiation passes through another diaphragm before the K_α radiation is absorbed by a 12 µm thick nickel filter. The x-rays are then detected by a standard scintillation counter. The sample rotates by a consecutive step process such that the angle of incidence of the primary beam gradually increased between preset values. The detector is correspondingly moved around the sample at precisely double the angular velocity ensuring that at all time the diffraction angle (2θ) is twice the glancing angle (θ). The control of the sample position, detector position, and scintillation data are controlled and recorded by a PC where a real time display of the data can be viewed. Data collection times were typically of the order of 2-3 hours and the scanning range between 2θ = 5 ° - 60 ° for samples studied herein.

2.2.2 XRS Data Analysis

XRD patterns were analysed by comparison with already published literature studies and with spectra produced using the LAZY PULVERIX [1] pattern simulation facility on a PC. The LAZY program calculates allowed *hkl* planes and the corresponding two-theta positions and intensities from crystal structure parameters entered in an input file which then allows identification of peaks in the XRD pattern.

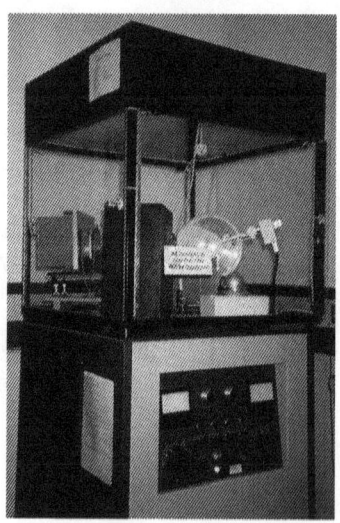

Figure 2.1a Photograph of the Phillips x-ray diffractometer.

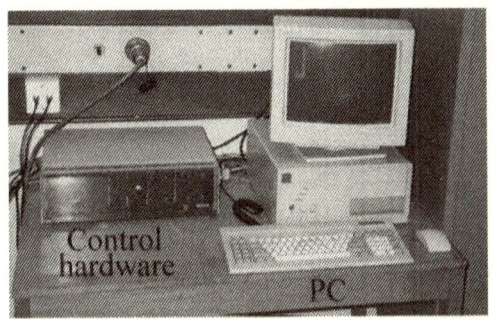

Figure 2.1b Photograph XRD the control hardware.

2.2.3 Theory of Diffraction

A crystal may be divided into layers by sets of planes passing through lattice points. Each of these sets of parallel planes may be described by three numbers such as 002 or 109, or in general *hkl*, where each *hkl* configuration represents a full set of planes running through the crystal structure separated by a perpendicular distance called the *d*-spacing. Each *hkl* in a crystal has a particular value for *d* and it is this quantity which may be related to angle and wavelength in a diffraction experiment through Bragg's Law.

$$2d(hkl)\sin\theta = n\lambda \quad \{2.1\}$$

where λ is the wavelength of the incident radiation and *n* is an integer. In theory a crystal should display diffraction from each of its lattice planes giving rise to an observed intensity in the resultant trace at 2θ values corresponding to all the planes. Intensity is not always observed at all possible positions however and this arises from the existence of reflection conditions or systematic absences that result from symmetry elements of the structure.

2.3 Scanning Electron Microscopy

The scanning electron microscope (SEM) is an instrument designed primarily for studying the surfaces of solids at high magnification. The benefits of the SEM over optical devices include the high magnifications obtainable, the greater resolution, and the improved depth of view. In the research described herein a Jeol 2010 scanning electron microscope (figure 2.2) was used to: (i) analyse the morphology of the samples prepared, (ii) determine superconductor grain sizes, and (iii) obtain values for the thickness of samples in order to calculate film void percentages.

2.3.1 SEM Operation

Images are formed in the SEM by a quite different mechanism from an optical microscope. No objective lens is used, but instead, images are built up point by point, in a way similar to that used in a television display. A fine, high-energy beam of electrons is focused to a point on the specimen. This causes the emission of electrons (with a wide spread of energies) from that point on the surface, and the emitted electrons are collected and amplified to give an electrical signal. If this signal is then used to modulate the intensity of a beam of electrons in a cathode-ray tube (CRT) display, one point of the image is formed on the CRT (see figure 2.3). To build up the complete image, the electron beam in the microscope is scanned over an area on the specimen surface (the pattern of scan is called a *raster*) while the beam in the CRT display is scanned over a geometrically similar raster. When the scanning beam reaches the part of the specimen corresponding to a darker region, very few electrons are emitted and the intensity of the synchronously scanned beam in the CRT is low; at a lighter region, the emitted intensity increases greatly and the CRT beam is made proportionately more intense. The image on the CRT is thus a map of the intensities of the electron emission from the specimen surface, just as the image in a metallurgical microscope is a map of the light reflected from the surface.

The magnification is given by the ratio of the side-lengths of the display and specimen rasters, and is normally variable from about $\times 20$ to $\times 100,000$. The resolution, however, is the more important quantity for any microscope, and with an ideal specimen it is, at best, equal to the diameter of the electron beam where it strikes the specimen surface (the beam in practice cannot be focused to a perfect point). In current high performance instruments, this can be as small as 5 nm. However, the resolution in the SEM depends critically upon the nature of the specimen. Resolutions of not much more than $\times 20,000$ were achievable for samples prepared for study.

(a)

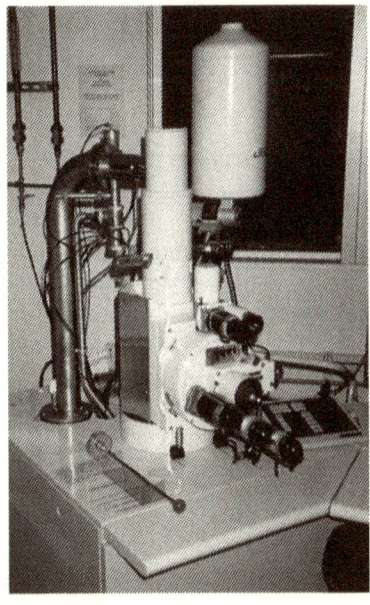

(b)

Figure 2.2 Photograph of the Joel 2010 scanning electron microscope. (a) control and analysis hardware, (b) electron gun and sample chamber.

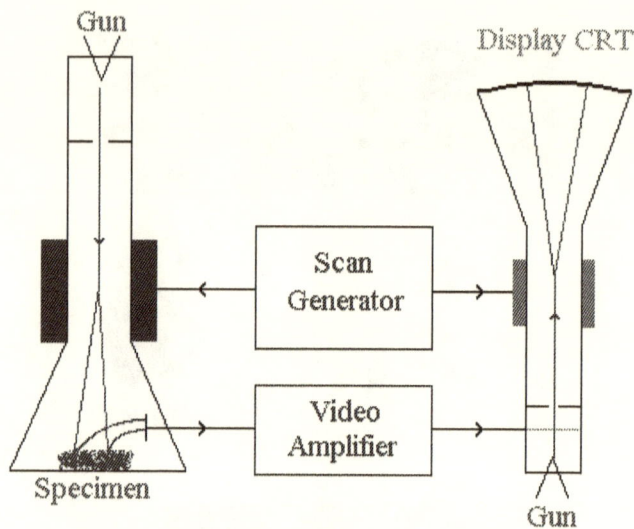

Figure 2.3 Schematic of a scanning electron microscope.

2.3.2 Sample Preparation

When analysing the morphology of the electrochemically produced films the samples were simply mounted on aluminium stubs with conductive carbon adhesive stickers. Typical acceleration voltages and magnifications used were 20 keV, and × 500 - × 1000. For powdered samples, small amounts of the powder were sprinkled over the aluminium stub with the conductive stickers. Care was taken to ensure that no loose material was left on the sample stub. Thin layers of gold were then evaporated onto the samples to improve the conduction of the samples so that the electrons produced by the SEM had a path to flow through.

The thickness' of electrodeposited films were determined by mounting a small section of the sample vertically in epoxy within a slit cut into a sample stub (see figure 2.4). Again, the samples were coated with a fine layer of gold to enhance conductivity. Sometimes thickness' were determined by simply mounting a sample onto a stub and then, when in the SEM, tilting the sample through 90° to view the sample cross section.

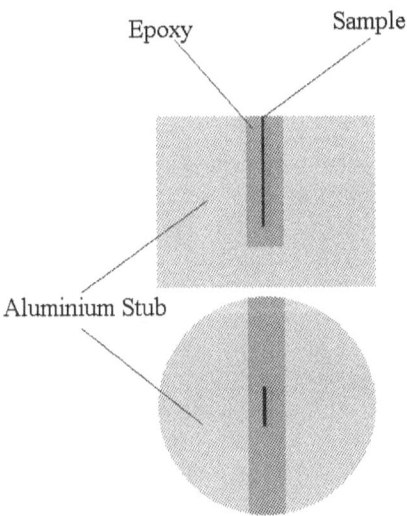

Figure 2.4 Sample mounting used in determining film thickness.

2.4 Energy Dispersive Spectroscopy

When the incident electron beam strikes the specimen, some of the electron energy is used in generating X-rays. The energies of the emitted X-rays are characteristic of the atoms present in the material. A solid state detector was used to collect the X-rays. The radiated photons were collected for a specific time period (100 s) and software used to determine the relative abundance's of the different atoms in the specimen, with an accuracy of ~ 10 %, by comparing the number of photons collected at particular energies, i.e. the intensity of the spectral lines. The characteristic lines, K_α, K_β, L_α, and so on are produced when electrons ejected from the K, L and other atomic shells in the target element are replaced by electrons from a higher-energy shell. The decrease in energy of the electron falling into a lower energy level is compensated for by the emission of a photon of appropriate frequency, v. The energy decrease and hence the photon frequency is characteristic of the element concerned, as expressed in *Moseley's law*:

$$\sqrt{v} = Z - C \qquad \{2.2\}$$

where Z is the atomic number of the target element and C is a constant specific to each characteristic line. This type of analysis is known as *energy dispersive spectroscopy* (EDS) and was used extensively in this work to determine the composition of the electrodeposited films. EDS was very useful in quantifying the level of reproducibility in the electrochemical process.

Problems arose when analysis of the Tl-Pb-Sr-Ca-Cu films were performed. The spectral lines obtained from the Tl and Pb in the samples were so close together that the software was unable to resolve them to give a useful value for their relative atomic abundances. Normally, the range of energies of the collected x-rays was set to 0 - 10 keV. However, when analysing the films containing Tl *and* Pb a range of 0 - 20 keV was applied. This is because that at higher energies there exist further spectral lines for these two materials that are not so close to each other and are therefore resolvable by the software accompanying the Jeol 2010 SEM. When quantifying the specimen composition the user may set which spectral lines are used in the calculation performed by the software so as to achieve the most accurate analysis.

2.5 AC Susceptibility

The susceptibility equipment used for the purposes of this study was developed and operated by Steve Pinfold and Darko Bracanovic within the Department of Physics in 1996, and is used herein to determine the superconducting transition temperature of powdered, PIT, and electrodeposited samples.

2.5.1 *Experimental procedure*

A small piece/quantity of the sample under study is placed inside a Teflon sample holder and secured to the end of the a sample rod which was then lowered into the sample space so that the holder layed in between the primary coil (Figure 2.5). The phases of both the primary and secondary coils were then adjusted using a lock-in amplifier until they were both close to zero. This calibration was especially important if the signals detected were very small. When a small AC current of constant amplitude is applied to the sample there is a voltage induced in both coils. The primary and secondary coils were however wound in serial opposition to each other. Hence, the induced voltage from the two coils cancel when the susceptibility of their cores is identical and a net signal, directly proportional to the sample susceptibility is observed. During χ_{AC} experiments it was assumed that the quantity of sample used was small enough to neglect variations in magnetic field across the sample.

The sample temperature was varied from 300 K down to a temperature below the superconducting transition using a constant stream of helium gas originating from a small 17 litre liquid helium dewar that passed through the sample chamber, cooling the sample, before exiting into a recovery line. The helium gas stream flowed due to the action of a combined rotary-diffusion pump system, the power of which could be varied according to the speed of cooling required. A heater was used to heat the sample to ambient temperature. All data obtained were taken during the warming of the sample as opposed to the cooling of the sample, as it was found that better results could be achieved. This effect is due to the fact that the sample was larger than the silicon diode temperature sensing device and therefore, as it had a larger heat capacity, it took longer to cool. These lagging effects were not so pronounced on warming. The helium transfer line was evacuated and the unit was housed within a styroform container filled with liquid nitrogen, reducing any unnecessary loss of helium to the atmosphere.

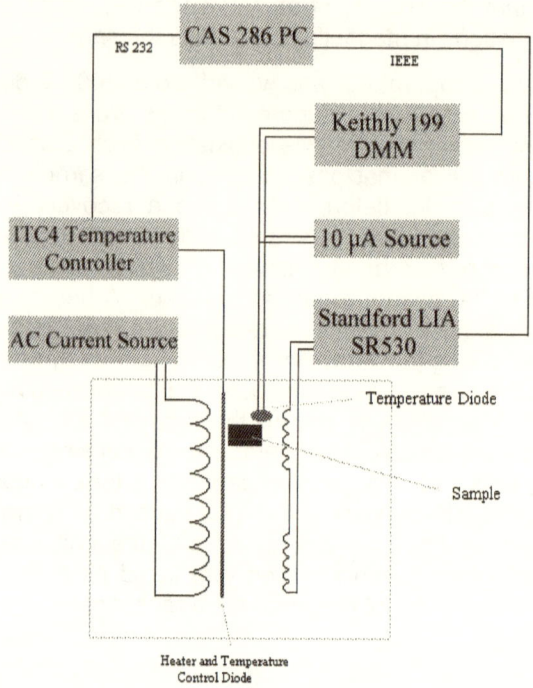

Figure 2.5 AC Susceptibility apparatus. The balance circuit has been omitted.

2.6 Vibrating Sample Magnetometry

The basic idea underlying vibrating sample magnetometry is similar to that of χ_{AC} measurements. In susceptibility measurements the voltage in the coils is varied and the change in the sample magnetisation monitored. In vibrating sample magnetometry, however, rather than the applied voltage being cycled, the position of the sample is varied and the back EMF, induced in a set of pick-up coils, measured. Magnetisation versus applied field measurements were performed on an Oxford Instruments 12 T vibrating sample magnetometer (VSM) (figure 2.6). These measurements were required to quantify the extent of the magnetic hysteresis within the superconducting samples prepared. From, the Bean model [2] the width of the hysteresis can be related to the *intrinsic critical current density*, sometimes also referred to as the *magnetic critical current density*, $J_{C,m}$, or the *intra-grain critical current density*. These measurements were performed on both PIT and electrochemically prepared samples. The samples were mounted flat so that the applied field was perpendicular to the plane of the sample or parallel to the *c-axis*.

Figure 2.6 A photograph of the Oxford Instruments 12 T vibrating sample magnetometer.

2.6.1 VSM Operation

The instrument works by moving the sample up and down between a set of pick-up coils, in this case at 60 Hz with an amplitude of 2.0 mm. If the sample has a magnetic moment this induces a voltage in the pick-up coils, proportional to the moment. The 12 T VSM comprises an Oxford 12 T magnet in a long-hold cryostat, fitted with integral variable-temperature insert, on top of which is mounted the VSM head. Control of the system is by an 'intelligent' control unit, the instructions for which are fed in from a PC at the start of an experiment. The unit performs all the field control, data collection and storage operations, and at the end of the experiment, data can be transferred to the PC for permanent storage. The temperature was controlled by an Oxford ITC4, which is set from the PC.

2.7 Squid Magnetometer

The SQUID magnetometer was used for this work to determine the superconducting transition temperature of electrodeposited superconducting films. The transition was obtained by determining the magnetic moment of the film versus temperature. All measurements were performed in an applied magnetic field of 2 mT.

The Superconducting Quantum Interference Device (SQUID) magnetometer (figure 2.7) allows DC fields of up to 6 T to be applied to the sample, using a superconducting solenoid. The set-up was fully automated, allowing both temperature sweeps and field sweeps. The magnet was housed in a liquid helium bath insulated from the ambient temperature with super-insulation, a vacuum space and liquid nitrogen bath. The cryostat was surrounded by a mu-metal shield which reduced the ambient field within the helium reservoir to a few µT.

The magnetic signal from the sample was detected using pick-up coils located centrally in the bore of the superconducting magnet. The coils were fabricated from superconducting wire and wound in a second order gradiometer configuration. This configuration rejects the field from the magnet to typically 0.1 %. The coils were connected to the input of the SQUID to form a flux transformer. Since the SQUID measures relative changes in magnetic flux, the sample is moved through the pick-up coils, setting up a screening current in the flux transformer circuit to oppose the change in flux threading the pick-up coils. This current, which is proportional to the induced magnetic moment of the sample, is detected by the SQUID via the input coil. A voltage directly proportional to the signal and hence magnetic moment, is output from the SQUID.

The superconducting magnet has a persistent mode switch (superconducting connection) across the magnet terminals, allowing large currents to continue circulating in the magnet without the need of input from the power supply. To change the set field, the current equivalent to the field was first ramped into the magnet supply leads. The switch was then driven into the normal state, by a heater, allowing the power supply to adjust the current flowing through the magnet. A second heater was also used to drive the pick-up coil flux transformer normal whenever the current supplied to the magnet was changed, and was necessary to prevent large currents being established in the transformer as a result of flux change through the pick-up coils.

The sample was mounted on a copper/silver wire, contained in a quartz tube. This was connected to a sample rod and placed in the Variable Temperature Insert (VTI). The insert was cooled using a continuous flow of helium gas from the helium reservoir, with the flow rate controlled by a needle valve. The cooling effect of the gas was balanced by a heater, allowing a steady temperature, in the range 1.7 K < T < 300 K to be maintained around the sample. The heater output was set using a Lakeshore controller, with the temperature measured using two rhodium-iron thermometers situated close to the sample. Valves controlling the gas flow into the VTI allow the sample to be removed whilst the insert was cold.

Figure 2.7 A photograph of the SQUID magnetometer.

The sample rod was smoothly moved between the pick-up coils using a stepper motor situated above the cryostat. The distance moved (scanlength) could be varied and was typically of the order of 1 to 6 cm. The magnetic field produced was homogeneous (to within 1 %) over a scanlength of ≤ 4 cm. Even for shorter scans, the field experienced by the sample would change by a finite amount during the scan, resulting in the sample effectively being subject to a small hysteresis loop. For each data point a number of scans are averaged, with the background signal being subtracted by the control software.

2.8 Transport Critical Current Rig

The transport critical current density, $J_{C,t}$, of samples of PIT tape were performed on a basic home-made rig (figure 2.8). Sections of tape (~ 3.5 cm in length) were wired up so that there were four voltage taps and two current leads (figure 2.9). The sample was then cooled to 77 K in a liquid nitrogen dewar. The voltage was measured between two of the voltage taps and the current through the sample increased. When the voltage reading reached 1 µV cm^{-1} the value of the current was recorded as the sample critical current. Different combinations of the voltage taps were used to obtain an average value of current. The sample mounting was positioned between the poles of an electro-magnet and a magnetic field was applied in ~ 0.01 T steps. At each step the critical current was determined again. The critical current was determined for the range 0 to 0.11 T for a field applied both parallel and perpendicular to the sample under study. The critical current density was obtained by simply dividing the value of the critical current by the samples' cross-sectional area, as determined by SEM. Differences between the two field applications reflect the level of texturing in the sample, a desirable feature of superconducting tapes.

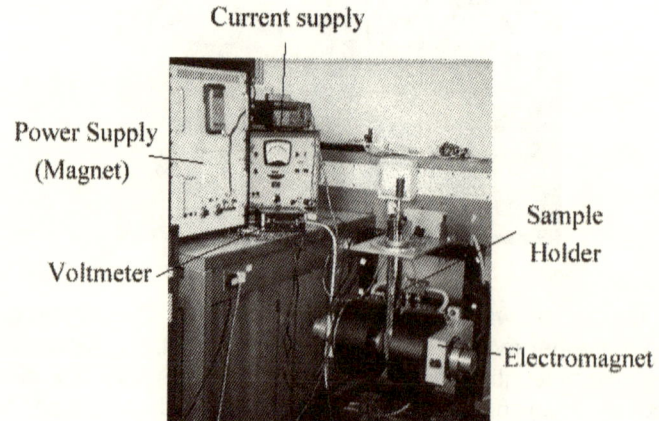

Figure 2.8 A photograph of the critical current rig.

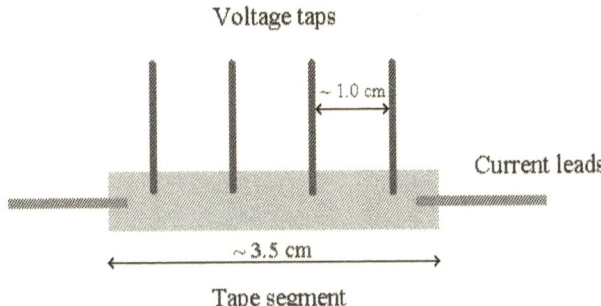

Figure 2.9 Schematic indicating connections made to tape segment for transport critical current characterisation.

2.9 Potentiostat

A potentiostat was used to perform all the electrochemical experiments involved in the electrodeposition of the superconductor precursor films. The potentiostat is a device for controlling the potential between the working electrode (cathode) and the reference electrode at a fixed and selected potential. There are several possible configurations, but the simplest, which is of little practical use but outlines the principles, is depicted in figure 2.10 where the cell has been approximated by a very simple equivalent circuit of a solution resistance in series with a double layer capacitance. It can be seen that the device is simply a voltage follower maintaining the output voltage between the reference electrode and the working electrode at the programming potential E_1. The working electrode, which is at ground potential, has a potential $-E_1$ relative to the reference electrode, so that the input voltage is inverted in the cell. Also in this arrangement there is no device for measuring the current through the cell, and the potentiostat is unable to apply more than a single potential profile at any one time.

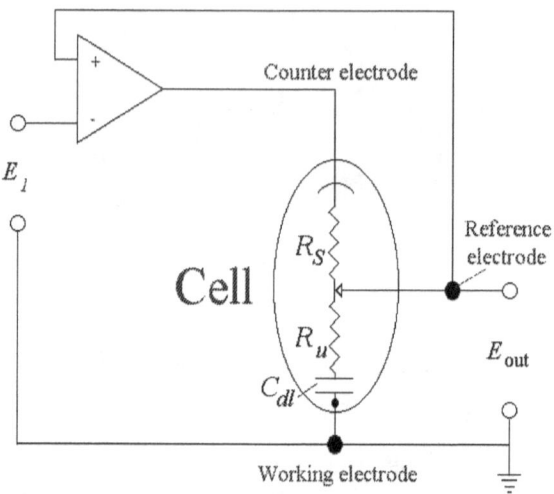

Figure 2.10 Circuit diagram for a simple potentiostat.

For this study an EG&G model 273A potentiostat was used (figure 2.11). This potentiostat is based on the above design but has many more features built-in. It is capable of applying a voltage versus time profile and measuring the resulting current passed, essential for cyclic voltammetry as well as chronoamperometry. The current is easily integrated to obtain the total net charge passed during an experiment, and the effects of iR_U drop can be compensated for by programming the potentiostat to perform a current interrupt (a particularly helpful feature). The hardware may be programmed from a PC or from a built-in key-pad, though the PC software is more flexible.

iR_U drop refers to the reduction in solution resistance close to the electrode/solution interface. The drop arises because of the decrease of electroactive species concentration due to electrochemical reduction.

Chapter 2 – Characterisation Techniques

Controlling PC

EG&G
Model 273 A

Figure 2.11 A photograph of EG&G 273A potentiostat and controlling PC.

2.10 Summary

In this chapter the experimental equipment and techniques for analysis of the following parameters has been described:

- (i) superconducting phase determination
- (ii) sample morphology
- (iii) sample composition
- (iv) superconducting transition temperature
- (v) transport critical current density
- (vi) magnetic critical current density
- (vii) magnetic hysteresis

Also, described briefly was the basic operation of the potentiostat used in fabricating the electrodeposited superconductor precursor materials.

The remainder of this thesis will be dedicated to reporting the research undertaken to fabricate superconducting tapes and films with the aim of a continuous production process kept in mind.

References

1. K. Yvon, W. Jeitschko, and E. Parthe, *J. Appl. Cryst.*, 10 (1977) 73.
2. M. Tinkham, *Introduction to Superconductivity*, McGraw-Hill, New York, 1975.

ns and Ag-Sheathed Tapes

3 Synthesis and Characterisation of Thallium-Based Powder and Ag-Sheathed Tapes

In this chapter the fabrication of thallium-based superconducting tapes via the already established PIT route is presented. The investigation was based around a newly discovered superconducting material. The successful preparation of single phase Tl-1223 powder with a composition of $(Tl_{0.6}Pb_{0.2}Bi_{0.2})$-$(Sr_{1.8}Ba_{0.2})Ca_2Cu_3O_{9+\delta}$, and the application of the powder for the preparation of Ag-sheathed tapes was researched. Presented here is a study of the microstructure of the as-prepared powder and the evolution of the prepared tapes' cores with temperature, as well as magnetic, superconducting, and transport properties.

3.1 Synthesis of the Thallium-Based Superconducting Powder

The main characteristics required from a superconducting material to be used in the manufacture of superconducting electrical devices, (especially power applications) capable of operating at liquid nitrogen temperatures, are that it needs to have a high transition temperature (§ 1.2), T_C, and also good current carrying properties in the presence of an applied magnetic field. For this reason the thallium-based compound $TlBa_2Ca_2Cu_3O_{9+\delta}$ (Tl-1223) [1] is currently being studied. Structurally, this compound has a single insulating Tl-O layer between the current carrying Cu-O layers (see figure 3.1). Compared to the double Tl-O layered compound $Tl_2Ba_2Ca_2Cu_3O_{10+\delta}$ (Tl-2223) (figure 3.2), the reduced spacing between Cu-O layers causes an increase in the coupling energy between the Cu-O planes. Kim et al. [2] have suggested that such an increase in the Josephson coupling energy inhibits the vortex lattice from breaking into pancakes, which are easily thermally activated. Thus, for temperatures at which thermal energies are significant, the current carrying properties of materials with increased coupling are less affected by applied magnetic fields. The strong coupling in Tl-1223 appears as a significant irreversible behaviour in the magnetisation loop at 77 K [3]. This property, combined with its T_C of 110 K, makes Tl-1223 a promising candidate material for practical applications.

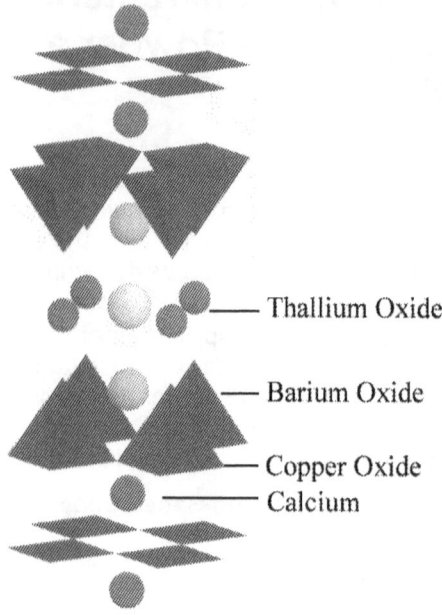

Figure 3.1 The atomic structure of Tl-1223.

Tl-1223 powder (partially reacted or fully reacted) with a morphology suitable for the fabrication of superconducting Ag-sheathed tape is not easily obtainable. It has been found that due to the high volatility of Tl the compounds are difficult to synthesise [4]. A starting composition of Tl:Ba:Ca:Cu = 1:2:2:3 yields the Tl-2223 phase as the predominant phase after sintering [1]. Further studies have revealed that the formation of the 1223 phase is via the sequence [5-7],

$$2212 \rightarrow 2223 \rightarrow 1223 \rightarrow 1234$$

indicating the difficulty in obtaining single phase $TlBa_2Ca_2Cu_3O_{9+\delta}$ superconducting powder. Ag-sheathed Tl-1223 tape has been manufactured with a ground powder of 'cauliflower' shaped $TlBa_2Ca_2Cu_3O_{9+\delta}$ multi-grains [8]. The grinding procedure, which was performed in order to reduce the particle size, resulted in the formation of an amorphous phase [9]. The fabricated tape showed very little texture and a transport critical current density ($J_{C,t}$) of 6.7×10^3 A cm^{-2} at 75 K and zero field [10].

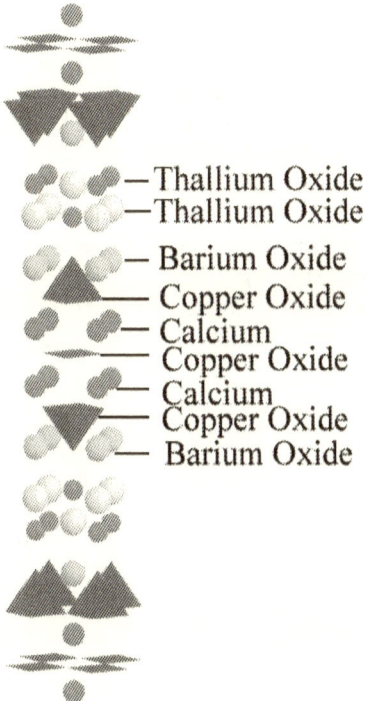

Figure 3.2 The atomic structure of Tl-2223.

Soon after the discovery of Tl-Ba-Ca-Cu-O high temperature superconductors, Subramanian et al. [11] identified $(Tl_{0.5}Pb_{0.5})$-$Sr_2Ca_2Cu_3O_{9+\delta}$ and showed that it was isostructural to the $TlBa_2Ca_2Cu_3O_{9+\delta}$ phase and with a similar T_C. When Pb^{4+} partially replaced the Tl^{3+} sites, the Tl-1223 phase is stabilised [12]. Furthermore, it was found that when Bi or Pb are partially substituted for Tl, and Sr partially or entirely substituted for Ba, nearly single phase Tl-1223 compounds can be obtained [13-17]. Torii et al. found that the amount of Bi included to partially substitute Tl could vary over a broad range without affecting T_C [14]. Liu et al. reported a simultaneous partial substitution of Tl with Bi and Pb, and Sr for Ba in the form of $(Tl_{0.6}Pb_{0.2}Bi_{0.2})$-$(Sr_{1.8}Ba_{0.2})Ca_2Cu_3O_{9+\delta}$ which resulted in an enhancement of T_C to 122 K and accelerated formation of the Tl-1223 phase [18].

3.2 The Manufacture of Thallium-Based Superconducting Tapes via the PIT* Technique

3.2.1 Background

To date, most developments of Ag-sheathed Tl-1223 tape have been made from $Tl(Ba,Sr)_2Ca_2Cu_3O_{9+\delta}$ with partial Pb and/or Bi substitution for Tl (table 3.1). Even though the transport critical current densities of the Tl-1223 tapes can be maintained at nearly 2×10^4 A cm^{-2} at 77 K and zero field, they decrease rapidly with increasing applied magnetic field. This rapid decrease in critical current density can be attributed to the weak inter-grain connectivity. As the intra-granular critical current density indicates, this material is very promising for 77 K operation in high magnetic fields if the connectivity between the grains can be improved.

Material	T_C (K)	$J_{C,m}$(77 K, 1 T) (A cm^{-2})	$J_{C,t}$(77 K, 0 T) (A cm^{-2})	Ref.
$(Tl_{0.5}Pb_{0.5})Sr_2Ca_2Cu_3O_{9+\delta}$	120	1.24×10^5	4.5×10^3	[25,26]
$(Tl_{0.6}Pb_{0.2}Bi_{0.2})(Sr_{1.8}Ba_{0.2})Ca_2Cu_3O_{9+\delta}$	110	6.5×10^4	5.6×10^3	This work
$TlBa_2Ca_2Cu_3O_{9+\delta}$	110	-	6.2×10^3	[8,10]
$(Tl_{0.5}Pb_{0.5})(Sr_{0.8}Ba_{0.2})_2Ca_2Cu_3O_{9+\delta}$	118	4.3×10^4	1.8×10^4	[23,24]
$(Tl,Bi)Sr_2Ca_2Cu_3O_{9+\delta}$	113	-	1.45×10^4	[21,22]
$(Tl,Pb)(Sr,Ba)_2Ca_2Cu_3O_{9+\delta}$	114	4.0×10^4	1.48×10^4	[17,20]
$(Tl_{0.5}Pb_{0.5})(Sr_{0.8}Ba_{0.2})_2Ca_2Cu_3O_{9+\delta}$	-	2.0×10^4	1.9×10^4	[19]

Table 3.1 Main parameters of selected Tl-1223 tapes.

3.2.2 Synthesis and Characterisation of (Tl,Pb,Bi)-1223 Powder

Fully reacted $(Tl_{0.6}Pb_{0.2}Bi_{0.2})(Sr_{1.8}Ba_{0.2})Ca_2Cu_3O_{9+\delta}$ powder was prepared by reacting Tl_2O_3, PbO, and Bi_2O_3 with a Sr-Ba-Ca-Cu-O precursor powder (figure 3.3). High purity $SrCO_3$, $BaCO_3$, $CaCO_3$, and CuO powders (Aldrich), in the appropriate proportions, were mixed in an agate mortar, containing acetone, for approximately thirty minutes to ensure a well mixed powder. The powder was then left for another 30 minutes for the

---

* PIT means powder-in-tube

Chapter 3: Synthesis and Characterisation of ....

remaining acetone to evaporate. The resulting mixture was then reacted for 16 hours at 920 °C in air. The reacted powder was ground further, and appropriate amounts of Tl_2O_3, PbO, and Bi_2O_3 powders were added. The powder was mixed once more, and then transferred to Au tubing and the tubing sealed at both ends. The reason for placing the powder in the tubing is to minimise Tl and Pb loss during the reacting process. The powder was then reacted, in flowing O_2, for 6 hours at 920 °C and cooled to room temperature at a rate of 2 °C min^{-1}. All heat treatment of samples was performed in a fume hood (figure 3.4) to ensure that Tl vapour did not escape into the laboratory atmosphere.

After the powder had been removed from the Au tubing and re-ground it was then characterised. The phase structure of the synthesised powder was determined by x-ray diffraction (XRD) analyses using a Philips diffractometer with $CuK\alpha_1$ (1.5406 Å) radiation (section 2.2). The composition of the powder was determined by inductive coupled plasma emission (ICP) spectroscopy[*]. The morphology of the powder was examined with the use of a Jeol 2010 scanning electron microscope (SEM) (section 2.3) with energy dispersive x-ray spectroscopy (section 2.4). The AC susceptibility of the powder was measured with a standard mutual inductance technique (section 2.5).

---

[*] ICP analysis was kindly performed by J. Guo of the University of Illinois, Urbana-Champaign, USA.

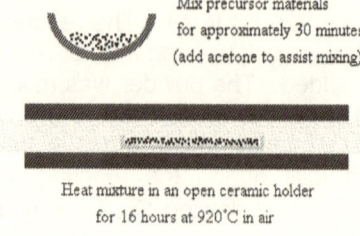

Mix precursor materials
for approximately 30 minutes
(add acetone to assist mixing)

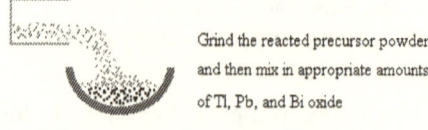

Heat mixture in an open ceramic holder
for 16 hours at 920°C in air

Grind the reacted precursor powder
and then mix in appropriate amounts
of Tl, Pb, and Bi oxide

 ← Oxygen

Heat the resultant mixture in a sealed
Au tube for 6 hours at 920°C
in an flowing oxygen atmosphere
and then furnace cool to room temperature
Finally, regrind powder and characterise

Figure 3.3 Pictorial representation depicting the experimental procedure for the preparation of (Tl,Pb,Bi)-1223 powder.

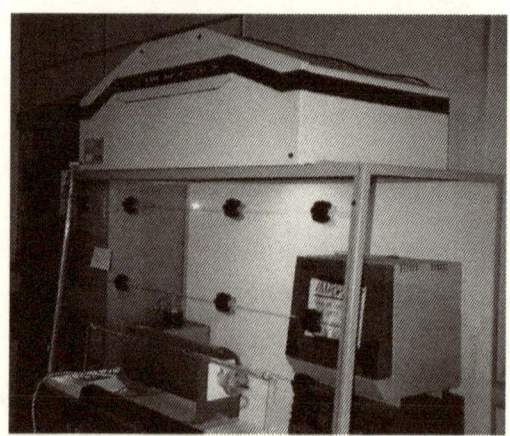

Figure 3.4 A photograph of the fume hood used to ensure that Tl vapour did not escape into the laboratory atmosphere. Inside the hood are the two furnaces for sample heat treatment.

3.2.3 Fabrication and Characterisation of (Tl,Pb,Bi)-1223 Tapes

Once the powder had been fully characterised it was then sintered for a further 1 hour at 850 °C in flowing O_2 to minimise the CO_2 content. It was then packed into an Ag tube ($\varnothing_{internal}$ = 3.0 mm, $\varnothing_{external}$ = 4.0 mm) (figure 3.5). Packing proceeded by pouring a measured amount of the powder into a tube. A plunger was then pushed into the tube and a 5 kg weight was placed on top for approximately 10 s. This process was repeated until a tube was full. Maximum packing density homogeneity was ensured along the length of the tubes by filling them in this way. The tubes, which were then sealed with In at both ends, were drawn into wires ($\varnothing_{external}$ = 1.187 mm), and then rolled to form tapes 2.47 mm wide and 0.184 mm thick. The resulting tapes were divided up into smaller sections, ~ 40 mm in length, and these sections were sintered at a range of temperatures (780 - 840 °C) for a different duration (2 - 12 hours) in air.

As with the powder samples, the sections of tape were characterised using several techniques. SEM was employed to view the tapes' cross sectional morphology, EDS to determine the tapes' core composition, and AC susceptibility to observe the superconducting transition temperature. Additionally, the intra-grain critical current densities ($J_{C,m}$), for different sections of tape as well as for the powder, were determined magnetically from measurements performed on a vibrating sample magnetometer (Oxford Instruments, model VSM 3001) (section 2.6). Finally, the field dependence of the transport critical current densities ($J_{C,t}$) of the tapes were determined using a four-probe method, the voltage criterion being 1 μV cm^{-1} (section 2.7).

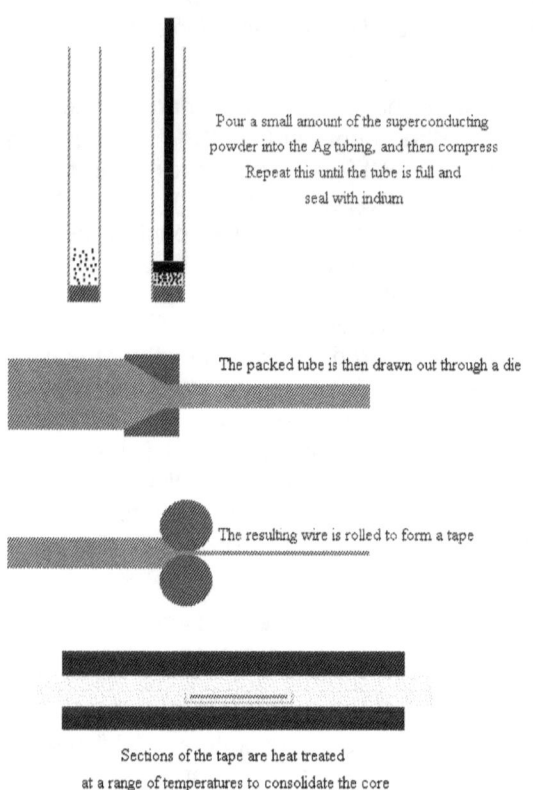

Figure 3.5 Schematic of Ag-sheathed tape manufacture.

3.2.4 Results and Discussion

Figure 3.6 shows the XRD spectrum of the as-synthesised powder. Almost all of the peaks were indexed to the Tl-1223 phase, indicating that the powder was very nearly monophasic. Moreover, results from ICP analysis concluded that the stoichiometric composition of the powder was given by $(Tl_{0.55}Pb_{0.19}Bi_{0.2})(Sr_{1.69}Ba_{0.19})Ca_2Cu_3O_{9+\delta}$, which is extremely close to the nominal composition. Additionally, ICP analysis indicated that there was ~5 parts per thousand of carbon present. The presence of carbon can be explained by the seemingly small amount of unreacted SrO. The SrO gradually absorbed CO_2 from the atmosphere and returned to its more stable form, $SrCO_3$. This fact has important consequences for tape making as the CO_2 can be released during post-roll heat treatment, causing the tape to bubble, or even burst.

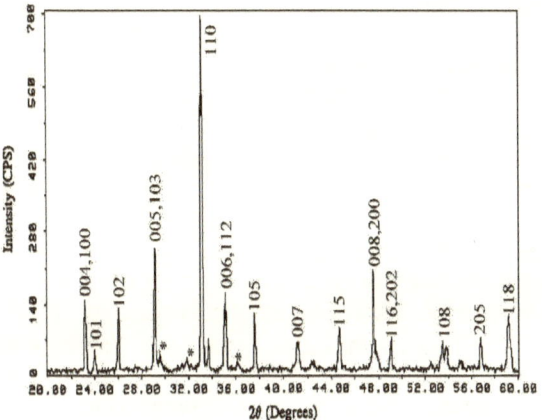

Figure 3.6 The XRD spectrum, with indexing, for fully reacted $(Tl_{0.6}Pb_{0.2}Bi_{0.2})(Sr_{1.8}Ba_{0.2})Ca_2Cu_3O_{9+\delta}$ powder.

The significant irreversible behaviour in an applied magnetic field that is intrinsic to the Tl-1223 superconducting structure (as mentioned above) can be seen in the hysteresis loops obtained from VSM measurements (figures 3.7a and 3.7b) at temperatures of 5 K and 77 K. At 77 K the as-prepared powder exhibited irreversible behaviour in applied magnetic fields of less than ~ 3 T. This result confirms the predictions of Kim et al.[2].

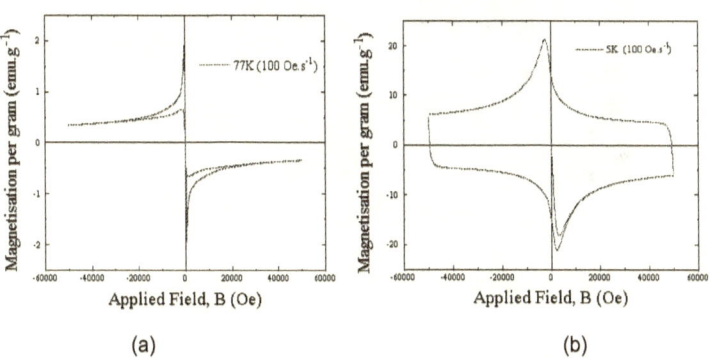

Figure 3.7 Hysteresis curves for as-synthesised $(Tl_{0.6}Pb_{0.2}Bi_{0.2})(Sr_{1.8}Ba_{0.2})Ca_2Cu_3O_{9+\delta}$ powder at (a) 5 K, and (b) 77 K.

Figure 3.8 is an SEM micrograph of the as-synthesised powder. The image indicates that the powder had a homogeneous platelet-type morphology, similar to $Bi_2Sr_2Ca_2Cu_3O_{10+\delta}$, with an average grain size of 4.5 μm and a maximum of 18 μm. The non-agglomerated morphology of this powder differs from that reported previously [9,10] in which as-synthesised $TlBa_2Ca_2Cu_3O_{9+\delta}$ powder was found to have a cauliflower-like structure. The cauliflower-like structure required further grinding in order to reach an average grain size suitable for PIT packing. The occurrence of platelet-type grains in superconducting powders is thought to contribute to the alignment of the grains in tape fabrication, and hence promote better transport properties. The occurrence of this characteristic in reacted $(Tl_{0.6}Pb_{0.2}Bi_{0.2})$-$(Sr_{1.8}Ba_{0.2})Ca_2Cu_3O_{9+\delta}$ indicates that this composition is a suitable candidate for Tl-1223 tape manufacture. The method by which the powder was produced was found to be very reproducible, another important requirement if consistent tape quality is to be obtained. Indeed, the fact that the phase formed and stabilised quickly is yet another positive attribute when considering the process of tape manufacture.

Figure 3.8 An SEM of the as-synthesised reacted $(Tl_{0.6}Pb_{0.2}Bi_{0.2})(Sr_{1.8}Ba_{0.2})Ca_2Cu_3O_{9+\delta}$ powder.

Figure 3.9 shows the normalised AC susceptibility of the powder. It can be clearly seen that the powder had an onset transition temperature of 110 K, similar to that observed elsewhere. It has been shown that this temperature can be raised to 122 K with further heat treatment [18]. The intra-grain critical current

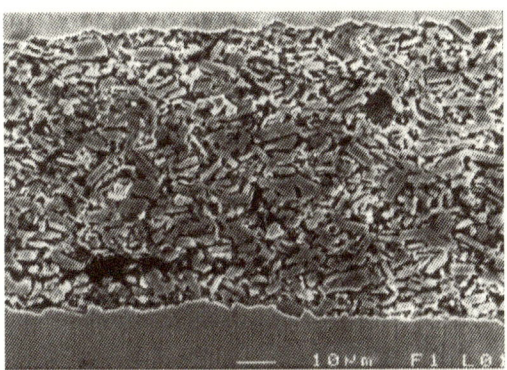

Figure 3.11 An SEM micrograph of the cross-section of a tape sintered at 820 °C for 2 hours in air and then quenched in air. The sample was etched before analysis.

From AC susceptibility measurements (figure 3.12) it was observed that the onset transition temperature of the tape was similar to the as-synthesised powder.

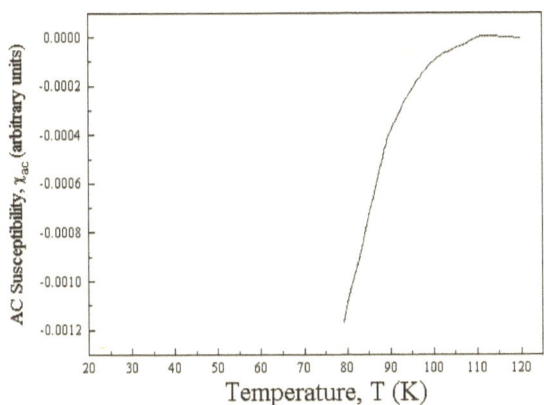

Figure 3.12 The AC susceptibility versus temperature for a section of tape.

Figure 3.13 represents the magnetic $J_{C,m}$ variation with applied field at 77 K for powder and tapes sintered for 2 hours at 780, 800, 820 and 840 °C, respectively. The $J_{C,m}$ values were determined by applying the Bean model (appendix A), where,

$$J_{C,m} = 30\Delta M/D \quad (A\ cm^{-2}) \qquad \{3.1\}$$

where ΔM (emu cm^{-3}) is the width of the magnetic hysteresis obtained from VSM measurements, and D (cm) is the average grain size (assuming low intergrain connectivity as the samples were quenched). The results show that there is no loss of intra-grain current carrying capabilities even with different sintering temperatures. The small differences in $J_{C,m}$ may be attributed to differences in the actual value of $J_{C,m}$, but are more likely to be inaccuracies in average grain size determination.

Figures 3.14a - 3.14c show the results from transport measurements performed on selected tapes at 77 K. Figure 3.14a represents the field dependence of the transport critical current density ($J_{C,t}$) for a section of tape sintered at 820 °C for 2 hours and then furnace cooled to room temperature. The $J_{C,t}$ at zero field was found to be approximately 4.0×10^3 A cm^{-2}. There were indications of texturing as the field dependence also depended on the applied field direction. Figure 3.14b shows the field dependence of a similar tape that had been annealed for a further 6 hours at 760 °C before being furnace cooled to room temperature. The $J_{C,t}$ had increased to 4.8×10^3 A cm^{-2}, which is not very impressive when considering the scatter that can be expected in these experiments. The field dependence, however, had also improved suggesting that better connectivity between the grains exist. The effect due to texturing is significantly more prominent which, as mentioned above, is unique to this system. Figure 3.14c shows the transport characteristics for a section of tape sintered for 12 hours at 800 °C in air and then furnace cooled to room temperature. The $J_{C,t(77K,\,0T)}$ is significantly higher (5.6×10^3 A cm^{-2}) but the field dependence, and evidence for texturing is poor due to the low sintering temperature.

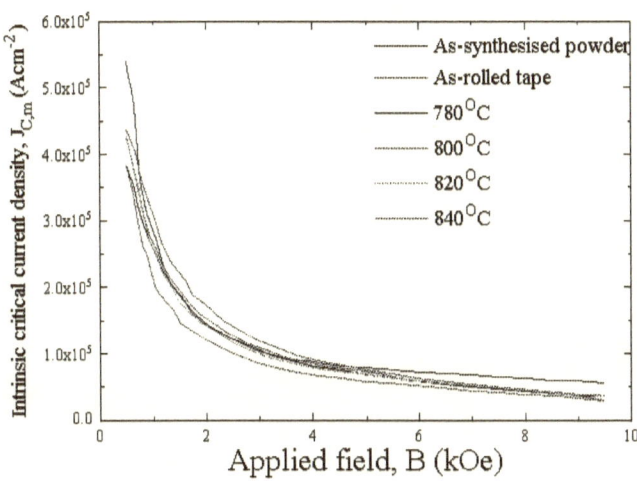

Figure 3.13 The intra-grain critical current density versus applied magnetic field for as-synthesised powder, as-rolled tape, and tapes sintered at 780, 800, 820, and 840 °C.

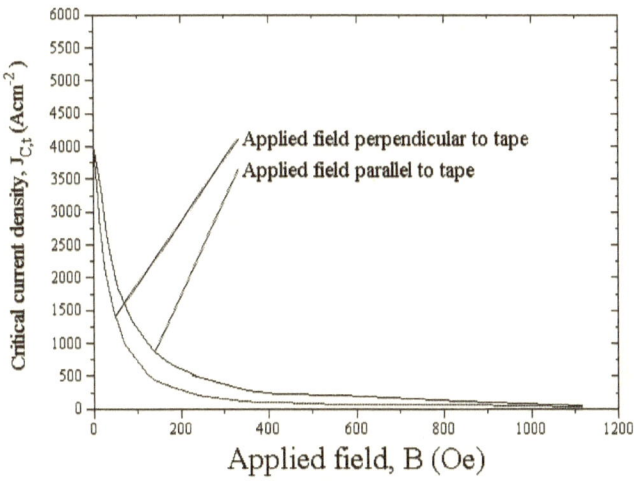

Figure 3.14a The transport critical current density, $J_{C,t}$, versus applied field, H, for a superconducting tape. The tape was sintered at 820 °C for 2 hours in air and then furnace cooled to room temperature

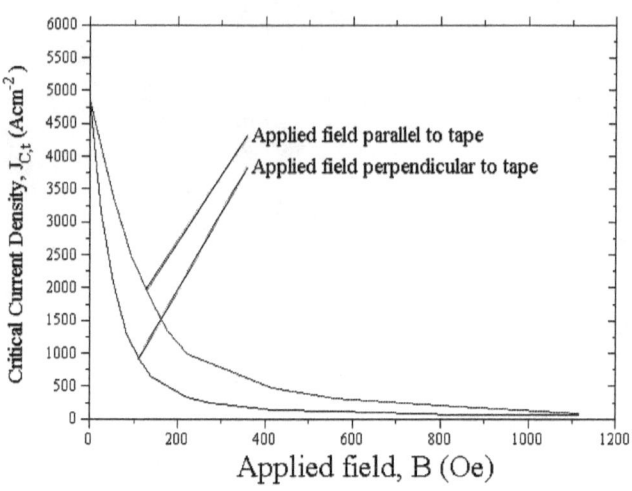

Figure 3.14b The transport critical current density, $J_{C,t}$, versus applied field, H, for a superconducting tape. The tape was sintered for 820 °C for 2 hours in air and then cooled to 750 °C for 6 hours and finally furnace cooled to room temperature.

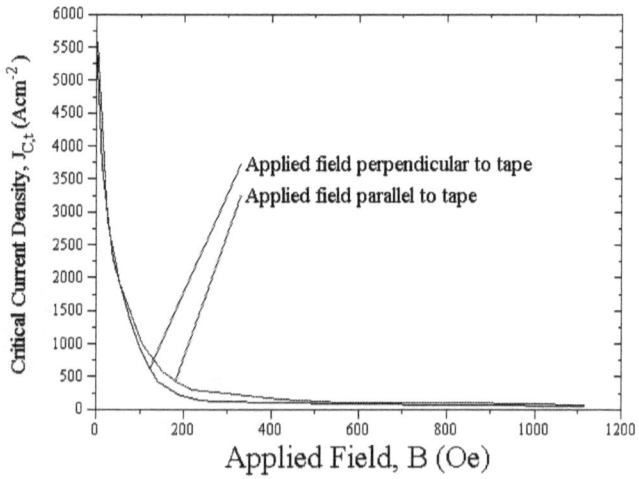

Figure 3.14c The transport critical current density, $J_{C,t}$, versus applied field, H, for a superconducting tape. The tape was sintered for 800 °C for 12 hours in air and then furnace cooled to room temperature.

3.2.5 Conclusions

In summary, a monophasic Tl-1223 compound has been synthesised with the stoichiometry $(Tl_{0.6}Pb_{0.2}Bi_{0.2})(Sr_{1.8}Ba_{0.2})Ca_2Cu_3O_{9+\delta}$ as suggested by Liu et al. [18]. The rapid formation and stabilisation of the phase was observed. The procedure was also found to be very reproducible. ICP analysis confirmed that the composition was very close to its nominal stoichiometry. ICP also indicated that small amounts of carbon were present (due to unreacted SrO) which had to be removed prior to packing into Ag tubes for tape fabrication. The morphology of the powder was found to be platelet-like and homogeneous, with intra-grain critical current densities of 6.3×10^7 A cm^{-2} at 5 K, 1 T, and 6.5×10^4 A cm^{-2} at 77 K, 1 T. The microstructure of Ag-sheathed tapes, produced using the powder, was investigated. Transport critical current density of 5.6×10^3 A cm^{-2} was obtained at 77 K for an applied field of 0 T. The tapes were found to be partially c-axis oriented which has not been previously observed in thallium-based PIT tapes. The platelet-like, homogeneous morphology, and high intra-grain current densities of the powder combined with the apparent texturing of the manufactured tapes suggest that this compound is very promising for applications. These characteristics promote high transport current densities but, as suggested by Ren et al. [27], inter-grain connectivity within the Tl-based tapes is a major deciding factor in the value of $J_{C,t}$ which is confirmed by results presented herein.

3.3 Further Work

The research presented in this chapter, which has formed the essence of a paper (reference [28]), has shown that the superconducting powder $(Tl_{0.6}Pb_{0.2}Bi_{0.2})(Sr_{1.8}Ba_{0.2})Ca_2Cu_3O_{9+\delta}$ has excellent morphology as well as a high intra-grain critical current density, as well as a high T_C. All these attributes are required for PIT tape candidature. The research has also shown that these properties are retained when the powder is used to fabricate tapes. The obvious problem with the material is the poor inter-grain connectivity. This problem must be overcome if the full potential of these tapes is to be realised, i.e. high current, high temperature, low response to applied fields. Work previously published [29-35] has shown that the inclusion of SnO_2 during the powder preparation may improve the inter-grain connectivity. This technique has been applied to Y-123, Bi-2223, and Tl-2223. The basic idea behind the technique is that the included SnO_2 does not react with the forming superconducting phase but coats the superconducting grains. Moreover, in the Tl-2223 phase system the inclusion of SnO_2 encourages phase formation that leads to increased superconducting fraction. Further work with $(Tl_{0.6}Pb_{0.2}Bi_{0.2})(Sr_{1.8}Ba_{0.2})Ca_2Cu_3O_{9+\delta}$ should first attempt to improve the inter-grain connectivity. This *may* include experiments measuring the effects of inclusions, such as SnO_2, upon tape transport properties.

Another approach to solving the problem of poor connectivity would be to use partially reacted superconducting powder so that the required phase is formed during post-roll heat treatment. Again difficulties may arise in the purity of the phase formed. Possibly a combination of partially reacted powder doped with Sn may achieve significant improvements to the connectivity and hence the current carrying properties of the developed tape.

Despite the success of the PIT technique in producing tapes and wires with promising superconducting properties the difficulties in fabricating long lengths ~ 100 km are clear. In particular, the problems involved in drawing such long lengths of wire such as fracturing of the silver sheath, and the inclusion of normal state material in the final tape, can result in long lengths of superconductor being inadequate for high current transmission. More recently it has been shown that pressing the wire into tapes improves the current carrying properties of the conductor [36], but the way by which pressing could by extended into a continuous process is unclear. For these reasons another fabrication technique was investigated - electrodeposition.

References

1   S. S. P. Parkin, V. Y. Lee, A. I. Nazzal, R. Savoy, R. Beyers, and S. J. La Placa, *Phys. Rev. Lett.*, 61 (1988) 750.

2   D. H. Kim, K. E. Gray, R. T. Kampwirth, J. C. Smith, D. S. Richeson, T. J. Marks, J. H. Kang, J. Talvacchio, and M. Eddy, *Physica C*, 177 (1991) 431.

3   T. Sasaoka, A. Nomoto, M. Seido, T. Doi, and T. Kamo, *Jap. J. Appl. Phys.*, 30 (1991) L1868.

4   W. L. Holstein, *Appl. Supercond.*, 2 (1994) 345.

5   R. Sugise, M. Hirabayashi, N. Terada, M. Jo, T. Shimomura, and H. Ihara, *Jap. J. Appl. Phys.*, 27 (1988) L1709.

6   R. Sugise and H. Ihara, *Jap. J. Appl. Phys.*, 28 (1988) 334.
7   E. Ruckenstein and C. T. Cheung, *J. Mater. Res.*, 4 (1989) 1116.
8   D. E. Peterson, P. G. Wahlbeck, M. P. Maley, J. O. Willis, P. J. Kung, J. Y. Coulter, K. V. Salazar, D. S. Phillips, J. F. Bingert, E. J. Peterson, and W. L. Hults, *Physica C*, 199 (1992) 161.
9   P. J. Kung, M. P. Maley, P. G. Wahlbeck, and D. E. Peterson, *J. Mater. Res.*, 8 (1993) 713.
10 P. J. Kung, P. G. Wahlbeck, M. E. McHenry, M. P. Maley, and D. E. Peterson, *Physica C*, 220 (1994) 310.
11 M. A. Subramanian, C. C. Torardi, J. Gopalakrishnan, P. L. Gai, J. C. Calabrese, T. R. Askew, R. B. Flippen, and A. W. Sleight, *Science*, 242 (1988) 249.
12 D. B. Kang, D. Jung, and M. -H. Whangbo, *Inorg. Chem.*, 29 (1990) 257.
13 M. Paranthaman, M. Foldeaki, R. Tello, and A. M. Hermann, *Physica C*, 219 (1994) 413.
14 Y. Torii, T. Kotani, H. Takei, and K. Tada, *Jap. J. Appl. Phys.*, 28 (1989) L1190.
15 O. Unoue, S. Adachi, and S. Kawashima, *Jap. J. Appl. Phys.*, 29 (1990) L763.
16 R. S. Liu, S. F. Hu, D. A. Jefferson, and P. P. Edwards, *Physica C*, 198 (1992) 318.
17 C. Martin, M. Huve, M. Hervieu, A. Maignan, C. Michel, and B. Raveau, *Physica C*, 201 (1992) 362.
18 R. S. Liu, S. F. Wu, D. S. Shy, S. F. Hu, and D. A. Jefferson, *Physica C*, 222 (1994) 278.
19 H. -T. Peng, Q. -Y. Peng, X. Y-. Lung, S. -H. Zhou, Z. -W. Qi, Y. -S. Wu, J. -P. Chen, Y. Zhoung, B. -S. Cui, J. -R. Fang, and G. -H. Cao, *Supercon. Sci. Tech.*, 6 (1993) 790.
20 Z. F. Ren and J. H. Wang, *Appl. Phys. Lett.*, 61 (1992) 1715.
21 Y. Torii, H. Kugai, H. Takei, and K. Tada, *Jap. J. Appl. Phys.*, 29 (1990) L952.
22 Y. Torii, H. Takei, K. Hasegawa, T. Kotani, and K. Tada, *Sumitomo Electric Tech. Rev.*, 36 (1993) 36.
23 T. Kamo, T. Doi. A. Soeta, T. Yuasa, N. Inoue. K. Aihara, and S. -P. Matsuda, *Appl. Phys. Lett.*, 59 (1991) 3186.

24 T. J. Doi, T. Nabatame, M. Okada, T. Yuasa, K. Tanaka, N. Inoue, A. Soeta, K. Aihara, T. Kamo, and S. -P. Matsuda, *Mat. Res. Soc. Symp. Proc.*, 235 (1992) (Pittsburgh, PA: Materials Research Society) p. 653.

25 B. A. Glowacki and S. P. Ashworth, *Physica C*, 200 (1992) 140.

26 26. R. S. Liu. D. N. Zheng, J. W. Loram, K. A. Mirza, A. M. Campbell, and P. P. Edwards, *Appl. Phys. Lett.*, 60 (1992) 1019.

27 Z. F. Ren, J. H. Wang, D. J. Miller, and K. C. Goretta, *Physica C*, 229 (1994) 137.

28 K. A. Richardson, S. Wu, D. Bracanovic, P. A. J. de Groot, M. K. Al-Mosawi, D. M. Ogborne, and M. T. Weller, *Supercond. Sci. Tech.*, 8 (1995) 238-244.

29 K. Osamura, N. Matsukura, Y. Kusumoto, S. Ochiai, B. Ni, and T. Matsushita, *Jpn. J. Appl. Phys.*, 29 (1990) L1621.

30 J. Shimoyama, J. Kase, S. Kondoh, E. Yanagisawa, T. Matsubara, M. Suzuki, and T. Morimoto, *Jpn. J. Appl. Phys.*, 29 (1990) L1999.

31 T. Lu, L. Shen, X. Sun, H. Shao, X. Jin, and N. Yang, *Mat. Res. Bull.*, 25 (1990) 315.

32 P. McGinn, W. Chen, N. Zhu, L. Tan, C. Varanasi, and S. Sengupta, *Appl. Phys. Lett.*, 59 (1991) 120.

33 A. S. Nash, K. C. Goretta, and R. B Poeppel, *Adv. Powder Metallurgy*, 2 (1991) 120.

34 H. M. Seyoum, J. M. Habib, L. H. Bennett, W. Wong-Ng, A. J. Shapiro, and L. J. Swartzendruber, *Supercond. Sci. Technol.*, 3 (1990) 616.

35 Z. Ren, M. Qi, and J. H. Wang, *Physica C*, 184 (1991) 24.

36 G. Papst, J. Kellers, and B. Gamble, to be published in *Superlattices and Microstructures*. Presented at the 5th World Congress in Budapest, July, 1996.

Chapter 4:

Electrochemistry and Electrodeposition of Superconductor Constituents

4 Electrochemistry and Electrodeposition of Superconductor Constituents

4.1 Introduction

Before attempting to deposit superconductor precursor films a study was made of the individual metal components of common superconductors. In this chapter an account is given of the investigation carried out on the electrochemistry and the electrodeposition of these constituent metals that was made in order to better understand the processes involved and their contribution to the electrochemical synthesis of superconductor precursor films. To start, the elementary theory of electrochemistry is discussed and phenomena such as: the double layer, the Helmholtz layers, and mass transport are introduced and explained. In order to extract information about the metals under consideration mathematical theory is developed to derive expressions for the concentration profile in the vicinity of the cathode and the current passed when limited by diffusion only. Furthermore, two electrochemical techniques applied to obtain fundamental information (i.e. the metallic cations' diffusion rates and their associated redox potentials) concerning the elements of interest are described. The chapter progresses to discuss the experimental parameters that were considered before the experimental procedure for superconductor precursor deposition could be designed. These include selection of a reference electrode, substrate selection and preparation, dry box electrochemistry, and iR_U drop. Finally, the results from the electrochemical investigations into Hg, Cu, Pb, Tl, Bi, Ba, Sr, and Ca are presented and discussed with particular reference to the fabrication of superconductor precursor films.

4.2 Theory of Electrochemistry

4.2.1 Basic Electrochemistry

Electrochemistry involves chemical phenomena associated with charge separation. The following section describes how a solvated metallic cation, M^+, moves from the bulk solution to the cathode surface where reduction may occur. Figure 4.1 is a schematic representing the basic experimental set-up using a standard three-electrode cell.

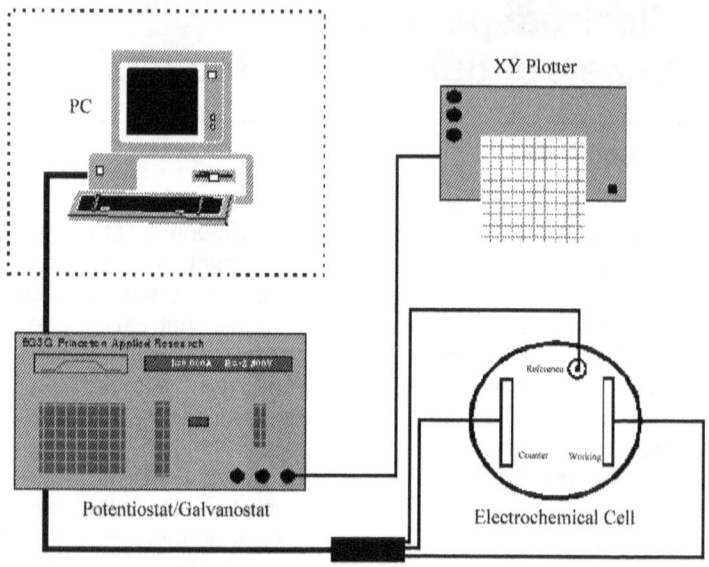

Figure 4.1 Basic experimental set-up.

A deposition solution is comprised of a solvent, sometimes containing a supporting electrolyte (see section 4.2.7.3), and dissolved ionic salts. Each solvated ion has an associated solvent atmosphere. The extent of the solvent atmosphere associated with a solvated ion (figure 4.2) depends on the size of the ion - a small ion has a greater charge density. The primary solvation shell of the ion is formed from solvent dipoles align with the ion (the motion of the dipoles is strongly restricted). The secondary solvation shell and the distorted solvent shell form an interfacial region between the primary solvation shell and the bulk solvent - the order of the solvent dipoles becoming progressively less the further away from the central ion we move. The ability of a solvent to dissolve the ionic salts is dependent upon its dielectric constant.

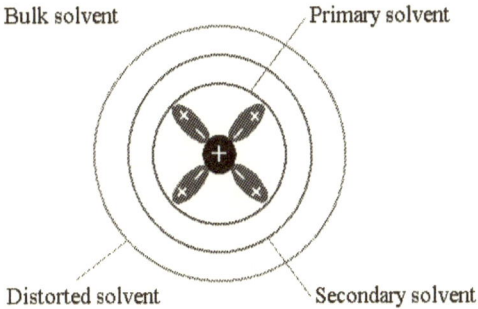

Figure 4.2 A metallic ion with its associated dipolar solvent atmosphere.

The solvated ions are free to move around the solution. Close to an electrode, however, asymmetric structure arises due of the existence of the metal/solution interface.

4.2.2 The Double Layer

The *interfacial region* arises as a result of the charge separation due to the interface between the electrode and the *electrolyte solution*. The part of the interfacial region that is within the electrode is known as the *space-charge region* [1-4], whereas the part of the interfacial region that is in the electrolytic solution is referred to as the *electrolyte double layer* [4-8]. In metals the space-charge region is very thin. Several models have been suggested that attempt to explain the structure of the double layer. As time has progressed these models have evolved to include successively more factors that contribute to the structure observed. The more common models are known as the: Helmholtz, Gouy-Chapman, Stern, and Grahame models, after their inventors.

The Helmholtz model [9] of 1879 considered the ordering of positive and negative charges, in a rigid manner, on the two sides of the interface giving rise to the double layer, or *compact layer*. The model assumed that the interactions did not act upon charges further within the solution. This meant that the system could be treated as a classical parallel plate capacitor problem with a measurable capacitance. Defects in this model include the neglect of interactions that occur further from the electrode then the first layer of adsorbed species, and also the omission of any dependence on electrolyte concentration.

Earlier this century Gouy (1910) and Chapman (1913) independently developed a double layer model in which they considered that the applied potential and electrolyte concentration both influenced the double layer capacity [10,11]. This meant that the double layer would not be compact, but of variable thickness, and the ions free to move. This is known as the *diffuse double layer*. Though a more realistic model than suggested by Helmholtz it is only valid at applied potentials close to the potential of zero charge†.

In 1924 Stern derived a model that combined both the Helmholtz and the Gouy-Chapman models [12]. This was achieved by applying the Helmholtz model for potentials far from the point of zero charge, and the Gouy-Chapman model for potentials close to the point of zero charge. This resulted in a compact layer in close proximity with the electrode and a diffuse layer that extended into the bulk solution. The boundary between the compact and diffuse layer was termed the outer Helmholtz plane. The Stern model distinguishes between adsorbed ions and those in the diffuse layer. However Grahame [13] extended this in 1947 by including specifically adsorbed ions: a specifically adsorbed ion loses its solvation, approaching closer to the electrode surface - besides this it can have the same, or opposite, charge as the electrode. Until the Grahame model (sometimes referred to as a *triple* layer model) was proposed the models treated the ions as rigid spheres, whereas Grahame took the step of including the chemical nature of the ions and the ion/electrode interactions.

More recently Bockris, Devanathan, and Müller (1963) have developed a more realistic model in which the dipolar nature of the solvent is taken into account [14]. In dipolar solvents such as water and dimethylsulfoxide (DMSO) there exists an interaction between the electrode and the dipoles. This interaction causes a predominance of solvent molecules near the interface. If we regard the electrode as a giant ion then the solvent molecules form its first solvation layer. This description of the double layer is depicted in figure 4.3.

---

† The potential of zero charge is the potential at which minimum double layer capacity occurs.

The culmination of the above models has resulted in the following description of the interfacial region. Closest to the electrode (assumed to be negatively charged) there will be a predominance of solvent molecules aligned perpendicular to the electrode surface, assuming a dipolar solvent. There will also be specifically adsorbed anions. These specifically adsorbed anions are anions that have lost their solvation and have bound to the electrode - the bonding generally being stronger than the coulombic bonding associated with the alignment of the solvent dipoles. The plane formed by the presence of specifically adsorbed anions is known as the inner Helmholtz plane (IHP). Further out from the electrode are the solvated cations. The cations are unable to reach the electrode, because of their solvation atmospheres, and form the outer Helmholtz plane (OHP). Outside the OHP is the diffuse layer which arises from the concentration gradient between the bulk of the solution and the cations forming the OHP. Figures 4.4 shows the variation of potential with distance from the electrode.

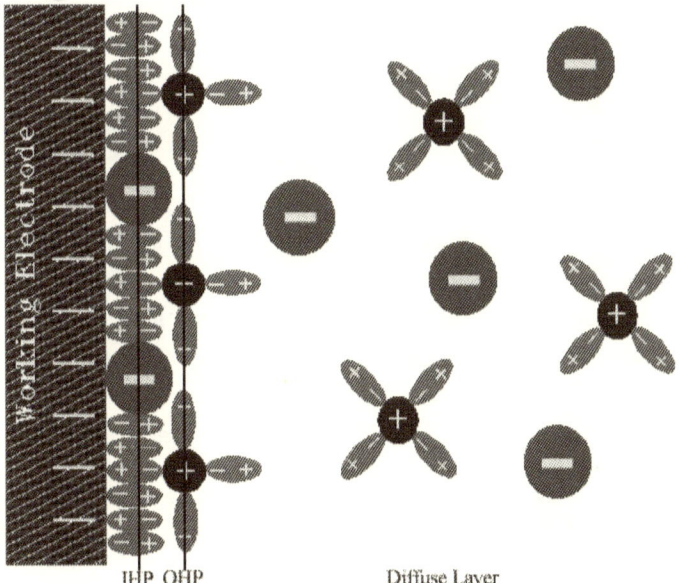

Figure 4.3 Schematic depicting the inner and outer Helmholtz planes, and the diffuse layer.

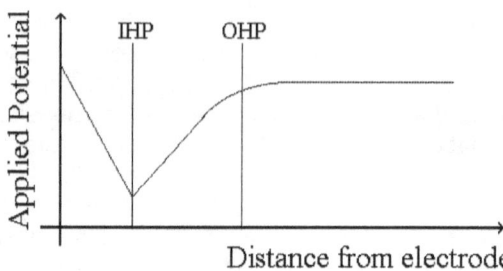

Figure 4.4 Potential versus distance from electrode for a planar electrode. Note that this is only one example of many possible forms for the potential distribution.

If a potential is applied to the cathode (working electrode), with respect to a reference electrode, that exceeds the redox potential of the metallic cations then electron transfer may occur. The probability of electron transfer increases the closer the cation is to the cathode, only becoming significant within the OHP. The probability that electron transfer will occur is dependent upon the solvent atmosphere, the tunnelling probability (which depends upon the distance between the electrode and the cation), and the size of the energy barrier to be surmounted. The rate of electron transfer is described by the following expression:

$$k_{et} = K_0 v_s k_{el} e^{\left(\frac{-\Delta G^*}{RT}\right)} \qquad 4.1\}$$

where K_0 is a constant; v_s represents the changing solvent atmosphere and therefore the cationic energy; k_e; is the tunnelling probability which has the form $\alpha \exp(-\beta r)$ where r is the electrode/cation separation; and ΔG^* is the activation energy (see figure 4.5).

Once electron transfer has occurred there is then a depletion of the initially present species which causes mass transfer from the solution by way of diffusion and migration. If a metal has been produced ($M^+ + e^- \rightarrow M^{(0)}$) then deposition may occur. The metal is referred to as an *adatom* which is insoluble in the solvent. The adatom travels across the electrode surface where it settles at a site with a preferential energy level. This is known as a *nucleation site*. All further deposition of adatoms is centred around such sites.

The above considers events at the working electrode, but what happens at the counter electrode (anode)? As the working electrode is connected to the counter electrode through a closed electronic circuit there must be conservation of charge. An electron transfers from the cathode to reduce a cation and therefore, an electron must be removed from the solution to the anode to ensure conservation of charge is maintained. The electron may be removed from an anion but could just as well be the result of solvent breakdown. The rate of oxidation is the same as the rate of reduction and to ensure that this is the case the electronics sets the potential at the anode to a value such that the current can be driven so that the potential at the cathode remains constant with respect to a reference electrode. The potential at the anode can, therefore, be very high compared to that at the cathode.

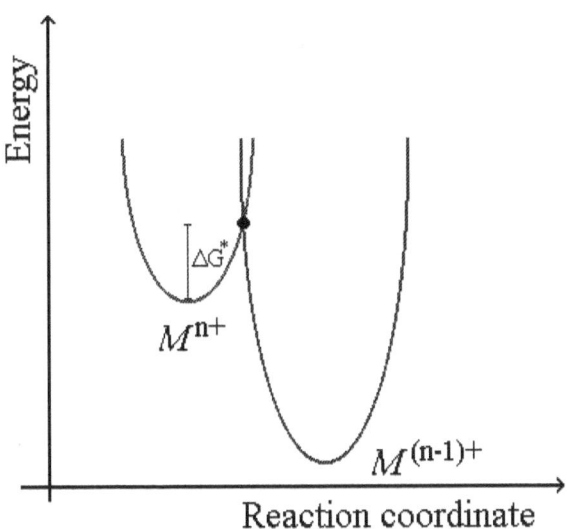

Figure 4.5 Schematic depicting the electron transfer reaction in reaction space.

4.2.3 Mass Transport

There are three mechanisms by which the cations in the bulk solution may reach the Helmholtz layer from the bulk solution: (i) diffusion, (ii) migration, and finally (iii) convection. The effect of the diffusive, migratory, and convective mechanisms upon ion movement may be represented by the following equation (in 1-dimension):

$$\frac{\partial c}{\partial t} = D\frac{\partial^2 c}{\partial x^2} - V_x \frac{\partial c}{\partial x} - Z_i c \frac{F}{RT}\frac{\partial \phi}{\partial x} \qquad \{4.2\}$$

where c is the concentration of the electroactive species (mol cm^{-3}), D is the diffusion rate (cm^2 s^{-1}), V_x is velocity of the ions (cm s^{-1}), x is the distance from the electrode (cm), Z_i is the charge on the ions, F is Faradays constant (96485 C mol^{-1}), R is the gas constant (8.31451 J K^{-1} mol^{-1}), T is the absolute temperature (K), and ϕ is the electrostatic potential (V). The first term in equation 4.2 is captures the effects of diffusion, the second term results from convective forces and, finally, the third term represents the effect on the ion movement from migration.

4.2.3.1 Diffusion

When the potential applied to the electrode is sufficiently large for reduction of the electroactive species to occur a region (close to the electrode) deficient in those ions immediately forms, and thence a concentration gradient. Diffusion is the movement of species down a concentration gradient. Therefore, within this diffusion layer (~ 10^{-2} cm thick) the concentration is dependent upon the distance from the electrode. The ionic flux flow is controlled by Fick's 2nd law (in 1-dimension):

$$\frac{\partial c}{\partial x} = D\frac{\partial^2 c}{\partial x^2} \qquad \{4.3\}$$

where D is known as the *diffusion constant*.

4.2.3.2 Convection

Convection is the movement of a species due to mechanical forces, such a thermal variations within the solution. It can be eliminated over short time scales ~ 10 s by performing experiments in a thermostatically controlled cell with no stirring or vibration. However, agitation of the solution is used frequently in electrochemical experiments to achieve a more uniform current flow.

4.2.3.3 Migration

There also exists a potential gradient within the electrolytic solution which causes charged species to migrate up or down the gradient in a direction dependent upon their charge (figure

4.6). The effect of migration, though, is easily removed by added a large excess of inert electrolyte to carry the charge.

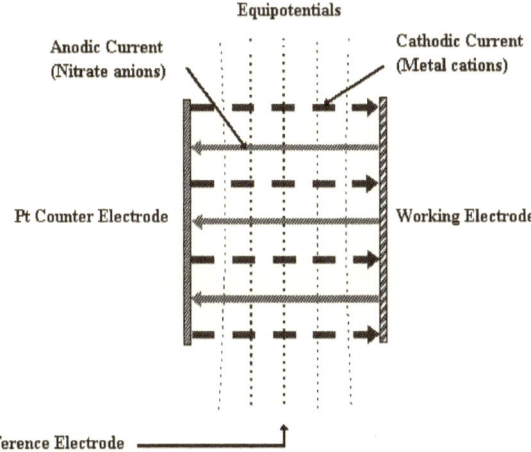

Figure 4.6 Charge migration due to a potential gradient.

4.2.4 Concentration Profile

This section begins by solving the diffusion equation to obtain an expression for the cationic concentration profile at the working electrode and subsequently the diffusion limited current. It is included as it clarifies some aspects of the electrochemical process. To obtain an expression for the concentration profile, $c(x,t)$, we must first solve the diffusion equation {4.3} with the following boundary conditions,

$t = 0$ $\qquad c_0 = c_\infty$ {4.4}

$t \geq 0$ $\qquad \lim_{x \to \infty} c = c_\infty$

$t \geq 0, x = 0$ $\qquad c_0 = 0$

where c_0 is the concentration at the electrode surface and c_∞ is the concentration in the bulk solution. To assist in the solving of the problem we introduce a dimensionless concentration,

$$\gamma = \frac{c - c_\infty}{c_\infty}$$ {4.5}

so,
$$\frac{\partial c}{\partial t} = D\left(\frac{\partial^2 c}{\partial x^2}\right) \rightarrow \frac{\partial \gamma}{\partial t} = D\left(\frac{\partial^2 \gamma}{\partial x^2}\right) \quad \{4.6\}$$

By transforming 4.5 with respect to t using the Laplace transform we obtain,

$$s\bar{\gamma} = D\left(\frac{\partial^2 \bar{\gamma}}{\partial x^2}\right) \rightarrow D\left(\frac{\partial^2 \bar{\gamma}}{\partial x^2}\right) - s\bar{\gamma} = 0 \quad \{4.7\}$$

This equation has the familiar solution:

$$\bar{\gamma} = A'(s)\exp\left[-\left(\frac{s}{D}\right)^{\frac{1}{2}} x\right] + B'(s)\exp\left[\left(\frac{s}{D}\right)^{\frac{1}{2}} x\right] \quad \{4.8\}$$

where $B'(s) = 0$ because as $x \rightarrow \infty$, $\bar{\gamma} \rightarrow 0$. Remember $x = 0$, $c_0 = 0$ and therefore $\gamma = -1$ and so $\bar{\gamma} = -(1/s)$. So for $x = 0$, $A'(s) = -(1/s)$. Therefore,

$$\bar{\gamma} = -\frac{1}{s}\exp\left[-\left(\frac{s}{D}\right)^{\frac{1}{2}} x\right] \rightarrow \gamma = -\mathrm{erfc}\left[\frac{x}{2\sqrt{Dt}}\right] \quad \{4.9\}$$

By substituting for c, c_∞ and rearranging we obtain an expression for $c(x)$,

$$c(x,t) = c_\infty \mathrm{erf}\left(\frac{x}{2\sqrt{Dt}}\right) \quad \{4.10\}$$

The surface (c,x,t) is depicted in figure 4.7. For small x an approximation for erf leads to the linear solution $c(x,t) \approx c_\infty x/2\sqrt{Dt}$.

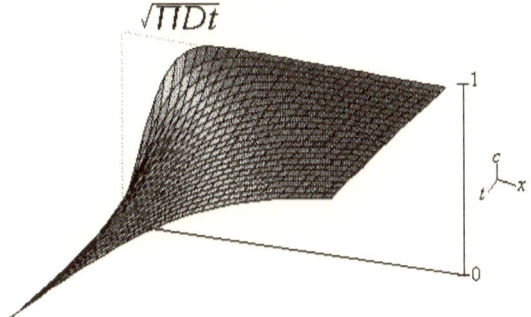

Figure 4.7 Surface plot of concentration, distance from electrode, and time for a planar electrode.

Given that $I = nFAJ \rightarrow nFAD(\partial c/\partial x)_{x=0}$ then we obtain an expression for the diffusion limited current, known as the *Cottrell equation*:

$$I(t) = I_d(t) = \frac{nFAD^{1/2}[O]_\infty}{\sqrt{\pi t}} \qquad \{4.11\}$$

The profile of equation of 4.11 is shown in figure 4.8. Also indicated is the concentration profile at small t and large t. From equation 4.10 it can be seen that $(\partial c/\partial x)_{x=0} = [O]_\infty / \sqrt{\pi D t}$ where $\sqrt{\pi D t}$ is the width of the diffuse layer.

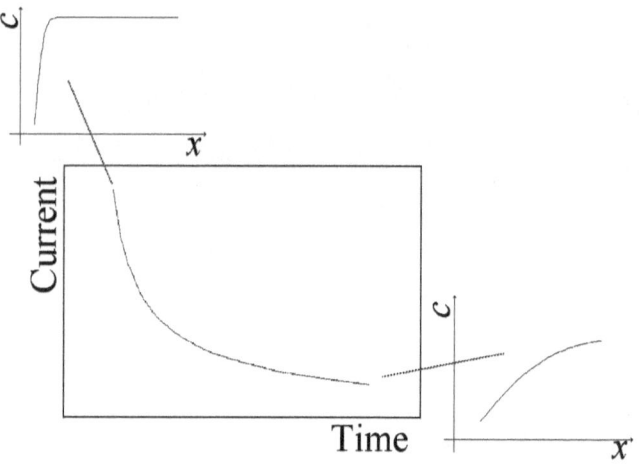

Figure 4.8 The diffusion limited current versus time profile according to the Cottrell equation. The insets depict the concentration profile for large and small time.

4.2.5 Cyclic Voltammetry

The most widely used technique in electrochemistry for studying electrode reactions is cyclic voltammetry, and is normally the first type of experiment performed in order gain information about the system under investigation because of the simplicity of the approach. The method can be used to investigate reaction kinetics, reaction rates, and in determining the species present in the solution under study. For the purposes of this work cyclic voltammetry was employed to determine the redox potentials, on Ag, of the relevant species dissolved in dimethylsulfoxide (DMSO). Additionally, values of the respective diffusion rates were determined.

The basis of cyclic voltammetry is the application of a time-varying potential to the working electrode. Whether a reduction or an oxidation of the species occurs (faradaic reactions) depends upon the range of the applied potential with respect to the redox potentials of the electroactive species. The total current due to the applied potential will also include a capacitive contribution from the charging of the double layer. Figure 4.9 shows the variation of the applied potential used for cyclic voltammetric analysis. The important parameters involved in this method are:

1. the initial potential, $E_{initial}$
2. the initial sweep direction, i.e. positive or negative
3. the rate of change of the applied potential, v
4. the maximum potential, E_{max}
5. the minimum potential, E_{min}
6. the final potential, E_{final}.

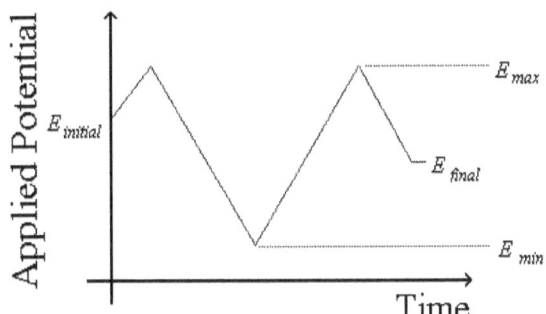

Figure 4.9 Applied potential versus time profile used in voltammetric analysis.

When the applied potential exceeds the redox potential of the electroactive species a faradaic current, $I_{faradaic}$, is produced, which manifests itself as a peak in the current versus applied potential profile. The magnitude of this current can be used to calculate the diffusion constant, D, for the species under observation given certain assumptions about the type of reaction and the number of electrons involved. In addition there is a capacitive contribution, $I_{capacitive}$, which must be taken into account when attempting to measure $I_{faradaic}$. This capacitive current is proportional to the sweep rate and therefore becomes more relevant at higher sweep rates. The total current is given by,

$$I = I_{capacitive} + I_{faradaic} = C_{double}\frac{dE}{dt} + I_{faradaic} = vC_{double} + I_{faradaic}$$

{4.12}

To find an expression to relate the diffusion constant, D, to $I_{faradaic}$, due to the simple electron transfer $O + ne^- \rightarrow R$ (where only O is initially present in the solution), we need to consider the kinetics and transport by diffusion of the electroactive species. Therefore it is necessary to solve the following equations simultaneously;

$$\frac{\partial [O]}{\partial t} = D_O \frac{\partial^2 [O]}{\partial x^2} \qquad \{4.13\}$$

$$\frac{\partial [R]}{\partial t} = D_R \frac{\partial^2 [R]}{\partial t^2} \qquad \{4.14\}$$

To solve 4.13 and 4.14 the following simple boundary conditions need to be applied:

$t = 0 \quad x = 0 \quad [O]_{x=0} = [O]_\infty \quad [R]_{x=0} = 0 \qquad \{4.15\}$

$t > 0 \quad x \to \infty \quad [O]_{x=0} \to [O]_\infty \quad [R] \to 0 \qquad \{4.16\}$

$t > 0 \quad x = 0 \quad D_O \left(\frac{\partial [O]}{\partial x} \right)_{x=0} + D_R \left(\frac{\partial [R]}{\partial x} \right)_{x=0} = 0 \qquad \{4.17\}$

$0 < t \leq \lambda \qquad E = E_{initial} - vt \qquad \{4.18\}$

$t > \lambda \qquad E = E_{initial} - vt + v(t - \lambda) \qquad \{4.19\}$

where λ is the value of t when the potential is inverted. Condition 4.15 simply states that at the beginning of the experiment the concentration of the electroactive species is constant throughout the solution. The next condition (4.16) implies that at large distances (> 30 μm) from the electrode surface the concentration of electroactive species remains unchanged from the initial value. Next, 4.17 the total flux of electroactive species at the electrode surface is zero, and finally, the last two conditions, 4.18 and 4.19, give the value of the applied potential at a particular moment in time. A fifth boundary condition is used to describe the kinetics of the electrode reaction. This is dependent on whether the system is reversible, irreversible, or quasi-reversible.

4.2.5.1 Reversible Systems

A reversible process exhibits equilibrium at the electrode surface. The equilibrium is described by the *Nernst equation*,

$$\frac{[O]_{OHP}}{[R]_{OHP}} = \exp\left[\frac{nF}{RT}(E - E^{\theta'})\right] \quad \{4.20\}$$

In order to solve 4.13 and 4.14 Laplace transforms must be applied. In this case the solution to the diffusion equations cannot be inverted analytically, and so must be inverted numerically. The result, after inversion, can be written as:

$$I = -nFA[O]_{\infty}(\pi D_o \sigma)^{\frac{1}{2}} \chi(\sigma t) \quad \{4.21\}$$

where

$$\sigma = \left(\frac{nF}{RT}\right)v \quad \{4.22\}$$

and

$$\sigma t = \frac{nF}{RT}(E_{initial} - E) \quad \{4.23\}$$

Figure 4.10 shows the curve of $\pi^{1/2}\chi(\sigma t)$ against $n(E - E^r_{1/2})$ where $E^r_{1/2}$ is the potential at which the current is equal to half the value of the peak current. The maximum value of $\pi^{1/2}\chi(\sigma t)$ occurs at (-28.50, 0.4463). This can be used to determine the diffusion coefficient of the reduced species. Firstly, it may be useful to attempt to understand the shape of the curve in Figure 4.10.

As the potential surpasses that at which the electroactive species reduces, the current rises as in a steady-state voltammogram. However, the reduction of the species causes a depletion of the electroactive species and thus the current begins to fall. The profile of the current thereafter is proportional to $t^{1/2}$ (as described by the Cottrell equation - equation 4.11), similar to the case when a potential step is applied to the system.

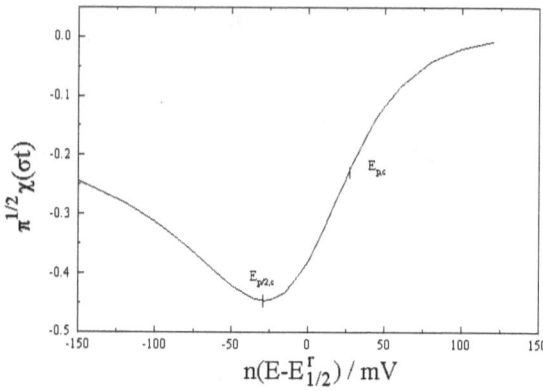

Figure 4.10 Normalised linear sweep voltammagram for a reversible reaction at a planar electrode. $E = E_{p/2}$ when $I = I_{p/2}$.

By substituting the maximum value in figure 4.10 into 4.23 we arrive at an expression for the peak current for a reversible reduction.

$$I_{p,c} = -2.66 \times 10^5 n^{3/2} AD_O^{1/2} [O]_\infty v^{1/2} \qquad \{4.24\}$$

with A measured in cm^2, D_O in cm^2 s^{-1}, $[O]_\infty$ in mol cm^{-3}, and v in V s^{-1}, at T = 303.16 K (30 °C). Therefore, from cyclic voltammetry we can not only determine the potential at which the electroactive species under investigation is reduced, but also the diffusion constant. The diffusion constant value calculated by this approach, in reality, does not give very accurate results, and is better determined using chronoamperometry (§ 4.2.6).

4.2.5.2 Irreversible and Quasi-reversible Systems

In irreversible reactions it is necessary to apply a higher potential than for reversible systems in order to overcome the activation barrier - the energy required for electron transfer to occur. Irreversible processes arise when the electron transfer rate is insufficient to maintain the equilibrium at the electrode/solution interface. In this situation equations 4.12 and 4.13 can again be applied but the boundary condition for $t > 0$, $x = 0$ is different from that used in the reversible process (equation 4.19). This boundary condition becomes,

Chapter 4: Electrochemistry and Electrodeposition.....

$$D_O \frac{\partial [O]_{x=0}}{\partial x} = k_0 e^{\frac{-\alpha_C nF}{RF}(E_{initial} - E^{\theta'})} e^{\frac{\alpha_C nFvt}{RT}} [O]_{x=0} \quad \{4.25\}$$

where α_C is the electrochemical charge transfer coefficient for a cathodic reaction, and k_0 is the rate constant (cm s^{-1}). By solving the differential equations again, we obtain the value of the peak current for an irreversible process to be (at 30 °C).

$$I_{p,c} = -2.96 \times 10^5 n^{3/2} \alpha_C^{1/2} [O]_\infty A D_O^{1/2} v^{1/2} \quad \{4.26\}$$

Other features of a totally irreversible process are that there is no oxidation peak in the cyclic voltammogram, and the reduction peak position is dependent upon the sweep rate - moving more negative with increasing sweep rate.

It is possible for a process to be reversible at low sweep rates and irreversible at higher sweep rates. The transition region between reversible and irreversible is referred to as *quasi-reversible*. As with the irreversible reaction the peak current increases with the square root of the sweep rate but, unlike the irreversible situation, is not proportional to it. Also the reduction peak moves more negative with increasing sweep rate.

4.2.6   Chronoamperometry

A step in the applied potential represents an instantaneous alteration to the electrochemical system. The rapid change can be analysed to make deductions concerning the electrode processes. Chronoamperometry is applied herein to yield measurements of the diffusion rates for the constituent superconductor components in the aim to obtain insight into the mechanisms of the electrodeposition of the superconductor precursor materials. A solution containing O initially and a large amount of supporting electrolyte, to ensure diffusion limited current, will produce a current whose magnitude is expressed by the Cottrell equation {4.10} when a potential is applied which is large enough to cause reduction of the species at a mass transport limited rate. If the current is then plotted against $t^{1/2}$ then the gradient will be given by,

$$\text{Gradient} = \Delta = \frac{nFAD^{1/2}[O]_\infty}{\sqrt{\pi}} \quad \{4.27\}$$

which can then be transposed to obtain a value of the species diffusion rate,

$$D = \pi \left(\frac{\Delta}{nFA[O]_\infty} \right)^2 \quad \{4.28\}$$

4.2.7 Experimental Considerations

Like most experiments, experimental conditions have a very noticeable affect upon data obtained. To ensure reproducibility, and the correct interpretation of the data obtained, it is extremely important to plan the experiments carefully. This section deals with important factors such as reference electrode and solvent selection, as well as working electrode cleaning. Also discussed in detail is the procedure to mitigate the detrimental influence of water contamination.

4.2.7.1 Reference Electrode

The reference electrode, as its name suggests, is used as a reference potential against which all potential measurements can be compared - potentials can only be registered as differences with respect to a chosen reference value. Apart from the cyclic voltammetry experiments, the experiments herein are known as potentiostatic (fixed potential) experiments where the potential between the working electrode and the reference electrode is maintained constant by a potentiostat (see section 2.6), and any change in the potential applied to the cell appears across the working electrode/solution interface.

The standard (or normal) hydrogen electrode (SHE) is the most important reference electrode because it is the one used to define the standard electrode potential scale. Many types of reference electrodes have been developed for aqueous electrochemistry, but fewer for non-aqueous electrochemistry. Ag/AgNO$_3$ in DMSO has been used as a reference electrode in the literature for the electrochemical synthesis of superconductors. However, when using this type of reference electrode for more fundamental investigations, such as cyclic voltammetry, it was found that the reference electrode potential was not completely stable. This may be due to the fact that the couple used is photoreactive - it is affected by light. Attempts were made to coat the reference electrode housing with dark paint or tape, but the DMSO dissolved these coatings. Instead, a quasi-reference electrode in the form of a Ag wire in DMSO was employed (+0.80 versus SHE) which provided better stability. The advantage of a quasi-reference electrode, such as, platinum and silver wires, or mercury pools, is the low electrical resistance. However, it must be remembered that the reference electrode potential is temperature dependent.

4.2.7.2 Solvent Selection

The solvent is the medium in which the electroactive species is supported, and also in which all the electrochemistry is performed. There is no universal solvent but there are selection criteria that can be used to choose the most appropriate solvent system. The most important are: electrochemical inertness, electrical conductivity of the solvent system[†], good solvent power, chemical inertness, and convenient liquid range.

Electrochemical inertness requires that the solvent in use does not decompose in the range of potentials being applied. This is known *as solvent breakdown*. The range required for the studies performed herein was +0.4 V to -3.5 V (at 30 °C and against Ag psuedo reference) which is an unusually large potential range. The solvent system must also have a low electrical resistance in order to support passage of an electrical current - this may be achieved with the addition of a supporting electrolyte as many solvents are insulators. Good electrical conductance reduces the effects of iR_u drop. Needless to say, the solvent must be able to dissolve a range of substances at acceptable concentrations - meaning that the solvent must have a large dielectric constant (> 10). For the purposes of superconductor deposition the solvent must be able to have within it dissolved nitrate salts at amounts reaching 500 - 600 mM (solvent power). Finally, the solvent must not take part in the process by reacting with the electroactive material, and also remain liquid over a useful temperature range.

The dipolar solvent dimethylsulfoxide (DMSO) (figure 4.11) was chosen for this investigation because of its exceptional solvent properties. It is stable down to ~ -2.5 V (versus Ag psuedo

---

[†] The solvent system is the combination of the solvent and the supporting electrolyte.

reference)[‡]; has a very high dielectric constant of 46.7; a high solvent power; is relatively inert; and is liquid in the range 18.0 to 189.0 °C.

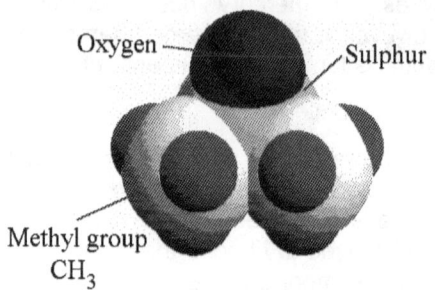

Figure 4.11 Molecular structure of dimethylsulfoxide.

4.2.7.3 Supporting Electrolyte and Complexing Agents

In an electrodeposition bath there can be, besides the cations for deposition, an electrolyte, known as a supporting electrolyte which transports most of the current, giving the solution a high conductivity. It is used in cyclic voltammetry and step potentiometry to overcome migrational effects (see 4.2.3.3). It can also, however, affect film morphology. In this work, for cyclic voltammetry and chronoamperometry analysis, tetrabutylammonium tetraflourophosphate and barium nitrate were used as electrolytes for experiments involving Sr, Ba, Ca and of Tl, Pb, Bi, Cu, Hg, respectively. The concentration of the supporting electrolyte is usually at least 100 times greater than that of the electroactive species when being added to overcome the effects of migration. The concentration of the electrolyte normally varies between 10 mM and 1000 mM, as opposed to 5 mM or less for the electroactive species.

Complexing agents are sometimes added. These additions change the redox potential of the cation to be deposited and also reduces the possibility of spontaneous chemical reactions occurring, giving the solution greater stability. No complexing agents were used in this study.

[‡] Despite some of the electrodeposition being performed at lower potentials (-3.00 V to -3.50 V) incorporation of sulphur into the films was extremely low.

4.2.7.4 Substrate Preparation

Silver was chosen as the working electrode for the deposition of superconductor precursors because of the low lattice mismatch with the desired superconducting phase which promotes c-axis alignment. Also, Ag is relatively cheap and has good flexibility, important when considering a continuous processing route. For deposition Ag foil was used as the working electrode and for more fundamental studies a Ag disc electrode was constructed (figure 4.12) with a diameter, $\varnothing$, ~ 2.8 mm.

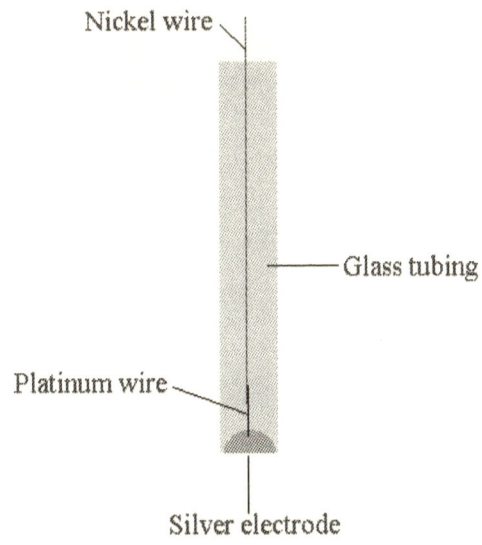

Figure 4.12 Ag disc electrode used for cyclic voltammetry.

Once the electrode had been constructed it had to be polished to obtain a smooth, brilliant surface which was free of physical defects on a length scale of ~ 0.3 µm. Substrates were polished with 0.3 µm alumina suspended in distilled water, by application and rubbing with cotton wool. Once the substrates had been polished for several minutes they were immersed in an ultrasonic bath containing water to remove any alumina fixed to the surface. After approximately five minutes in the bath the substrates were cleaned with cotton wool, and distilled water, and once again immersed in an ultrasonic bath to ensure maximum cleanliness. The substrates were then dried using cotton wool and placed in a hot cupboard (held at 100 °C) to ensure that they were completely dry and free from water that may contaminate the experiments.

4.2.7.5 Nitrate Salt Dehydration

The following section (4.2.7.6) details the reasons for minimising water contamination in the deposition solution. The nitrate salts used for the cyclic voltammetric studies of the superconductor constituent metals, and the superconductor precursor films, had to be completely dried before use in any experiments. Salts with a reasonably high melting temperature (see table 4.1) were dried at 200 °C in a box furnace for at least one week to evaporate off any contaminant water before introduction into the dry box. Metallic salts with relatively low melting points, in particular the calcium and copper salts, were stored over phosphorous pentoxide for approximately one month to minimise water content. The resulting salts, however, would still have contained trace amounts of water but this was not found to significantly affect the experiments performed. In particular, it is difficult to remove the final water molecule from the calcium and copper nitrates [15]. In these cases it is assumed that they still contain one water molecule per nitrate molecule.

Metallic Salt Name	Symbol	Melting Temperature (°C)
Copper Nitrate Hydrate	$Cu(NO_3)_2H_2O$	114
Thallium Nitrate	$TlNO_3$	430
Lead Nitrate	$Pb(NO_3)_2$	470
Bismuth Nitrate	$Bi(NO_3)_3$	-
Mercury Nitrate	$HgNO_3$	-
Strontium Nitrate	$Sr(NO_3)_2$	570
Barium Nitrate	$Ba(NO_3)_2$	592
Calcium Nitrate Hydrate	$Ca(NO_3)_2H_2O$	44

Table 4.1 The melting temperatures of relevant metallic salts (reference 16).

4.2.7.6 Dry Box Electrochemistry

Control of the atmospheric concentrations of oxygen and water is an essential requirement for the electrodeposition of the superconductor components if a high level of reproducibility is to be achieved, and accurate data obtained. The environmental control is easily attained but the time consumed in ensuring the appropriate conditions significantly increases the duration of the experimental preparation period. What are the effects of oxygen and water on the deposition process that necessitates the removal of these contaminants?

The solvent used in the electrochemical process, dimethylsulfoxide (DMSO), is anhydrous but quite hygroscopic and should therefore not be exposed to the laboratory atmosphere unnecessarily. The time it takes to pour or pipette from one container to another can be sufficient as to cause the absorption of significant amounts of water into the DMSO. As the metallic salts are also hygroscopic, this can lead to unacceptable concentrations of water being incorporated into the deposition bath. Problems with water reduction are created at high overpotentials applied to the working electrode. Additionally, water may form complexes with the metallic cations and hence affect their electrochemistry, e.g. their redox potential may change. The reduction of water ($2H_2O + 2e^- \rightarrow H_2 + 2OH^-$) causes hydrogen bubbling at the working electrode which interferes with the deposition process. Consequently, the film homogeneity, stoichiometry, process reproducibility (due to the variability of the water content), and ultimately the resulting superconducting properties are affected.

Oxygen also interferes with the process and its presence must therefore be minimised. The laboratory atmosphere contains about 20 % oxygen, which is a little heavier than air, and is dissolved appreciably ($\approx 10^{-4}$ M) in solutions open to the atmosphere. Oxygen is reduced at the working electrodes in two separate 2-electron steps or in one 4-electron step as follows,

$$O_2 + 2H^+ + 2e^- \rightarrow H_2O_2$$

$$H_2O_2 + 2H^+ + 2e^- \rightarrow 2H_2O$$

the 4-electron reduction being obtained by adding the two 2-electron steps together, i.e. $O_2 + 4H^+ + 4e^- \rightarrow 2H_2O$. The presence of these half-reactions contributes to the current measured at the electrode as well as creating water that affects the process in the manner described above. The protons required for the above reaction pathways to occur are likely to come from the auto ionisation of water: $H_2O \rightarrow H^+ + OH^-$. The oxygen may also oxidise the substrates' surface. How is the concentration of oxygen and water minimised?

Drying Agent	Residual Water per Litre of Air (mg)
$CuSO_4$	2.8
$CaCl_2$	1.5
$ZnCl_2$	1.0
NaOH	0.8
H_2SO_4 (95 %)	0.3
Silica Gel	0.03
$Mg(ClO_4)_2 2H_2O$	3.003
KOH	0.014
Al_2O_3	0.005
$CaSO_4$	0.005
H_2SO_4	0.003
$Mg(ClO_4)_2$	0.002
BaO	0.0007
P_2O_5	0.00002

Table 4.2 Comparative drying efficiency of drying agents (reference 17).

Figure 4.13 shows the type of dry box used. Essentially, it is a large dessicator with an argon atmosphere. Phosphorous pentoxide (P_2O_5), which is has the highest known drying efficiency of all commonly used drying agents (see table 4.2), is used to keep the atmosphere dry. Ultra pure argon gas, which is inert, was pumped into the box to keep the partial pressure of oxygen to a minimum. The components of the deposition bath were stored under these same conditions, and the deposition bath itself was prepared in the dry box. To minimise the oxygen content of the bath further, it was ultrasonically irradiated for approximately 5 minutes, and the box repeatedly purged (typically 8 - 10 times) with argon before the commencement of any electrochemical experiments.

Chapter 4: Electrochemistry and Electrodeposition.....

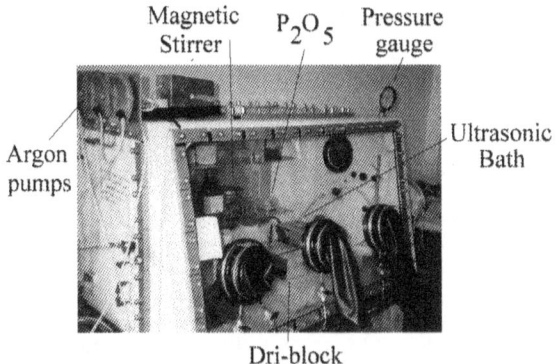

Figure 4.13 The dry box used to control the experimental atmosphere.

4.2.7.7 iR_U Drop

Inaccuracies in measurement may arise through an effect known as iR_U drop (see § 2.9). This effect is occurs because of the uncompensated solution resistance, R_U. This resistance means that when a potential, $E_{applied}$, is applied the actual potential applied, as in the potential applied to the working electrode with respect to the reference electrode, is given by,

$$E_{actual} = E_{applied} - iR_U \qquad \{4.25\}$$

This error can be limited by minimising R_U by placing the reference electrode as close as possible to the working electrode. The reference potential is monitored at the tip of a *Luggin capillary* (see figure 4.14) which is placed in the potential gradient by the current flowing between the counter and working electrodes.

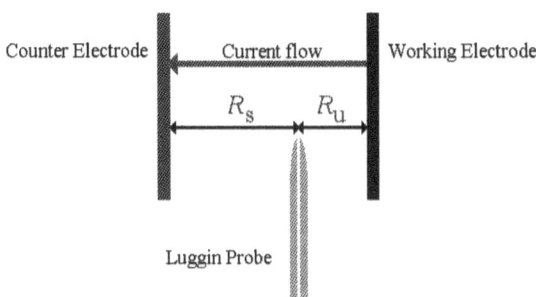

Figure 4.14 Schematic indicating how the problem of IR drop arises.

In experiments performed herein the potentiostat was sometimes set to *current interrupt* to compensate for iR_U drop. This meant that during the experiment the potential was switched for a brief period to monitor the drop in current. This drop was used to calculate, in real time, the iR_U drop. During such experiments the potential drop never exceeded 10 mV.

4.3 Electrochemistry of Superconductor Constituents

4.3.1 Experimental

The investigations into the properties of the individual superconductor precursor constituents were all performed at 30 °C in a dry box (§ 4.2.7.6). Care was taken to ensure minimal contamination from oxygen and water. All redox potentials were measured against a Ag psuedo reference, with a piece of Pt gauze as the counter. An Ag electrode with an area of ~ 0.06 cm^2 (§ 4.2.7.4) was used as the working electrode. The deposition solutions comprised of 5.0 mM of the metal salt to be studied, plus 500.0 mM of supporting electrolyte, both dissolved in DMSO. In the cases of mercury, copper, lead, thallium, and bismuth, barium nitrate was used as the supporting electrolyte. Whereas, in the cases of barium, strontium, and calcium, tetrabutylammonium tetraflouroborate ($C_{16}H_{36}BF_4N$) was used. Two types of experiments were performed in order to determine E_{red}, D_{CV}, and D_{SV} of the metals. The first type was simple cyclic voltammetry (§ 4.2.5), whereby the cathodic current was measured as a function of applied potential swept at a particular rate. To obtain further information of the electrochemical processes involved the sweep rate was varied to quantify changes in E_{red} against v. Secondly, step voltammetry was performed (in order to calculate D_{SV}).

4.3.2 Electrochemistry of Copper

Figure 4.15 shows the variation in the shape of cyclic voltammograms with sweep rate for 5.4 mM copper nitrate dissolved in DMSO. It is clear that the deposition of copper metal is a two stage process, the first stage being represented by the redox reaction: $Cu^{2+} + e^- \rightarrow Cu^+$, and the second stage being represented by $Cu^+ + e^- \rightarrow Cu^{(0)}$. From the figure the redox potentials for the first and second stage reactions are -0.08 V and -0.38 V, respectively. The oxidation peak of $Cu^{(0)}/Cu^+$ was also observed (displaying a peak current at ~ -0.18 V) indicating that material had been deposited. The beginning of the Cu^+/Cu^{2+} oxidation peak was also observed.

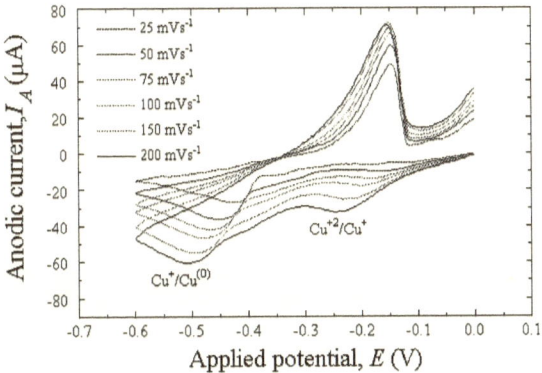

Figure 4.15 *The variation of cathodic current versus applied potential for different sweep rates for copper. The area of the electrode was 0.06 cm² and the solution comprised 5.40 mM $Cu(NO_3)_2H_2O$ and 500.0 mM of $Ba(NO_3)_2$ dissolved in DMSO. The solution was maintained at 30 °C during the experiment.*

Also to be noted from figure 4.15 is the shift in the peak current potential, E_{peak} with the sweep rate, v. This effect may be due to the kinetics of the process, to IR drop, or a combination of both. The theory of a reversible process demands that the distance between the peak reduction current potential and the peak oxidation current potential is $56.6/n$ mV. For Cu^+ this difference is ~ 280 mV indicating that the process is not a reversible one. The theory for an irreversible process predicts a shift in E_{peak} with v. The relationship can be described by the following equation:

$$E_{peak} = E^{\theta'} - \frac{RT}{\alpha_c n'F}\left[0.780 + \ln\frac{D_0^{1/2}}{k_0} + \frac{1}{2}\ln\frac{\alpha_c n'Fv}{RT}\right] \quad \{4.30\}$$

So if E_{peak} is plotted against $\ln(v)$ than a straight line should be obtained. Figure 4.16 shows this relationship and a reasonable straight line fit was obtained. The equation of the fit is given by $E_{peak} = -0.181 - 0.039\ln(19.14v)$. The gradient of the curve should be $-(1/2)(RT/\alpha_c n'F)$ which when calculated equals -0.026 ± 0.013 V if α_C is assumed to lie in the range $0.25 \rightarrow 0.75$. If the upper value of -0.039 V is compared to the experimental value of -0.039 V there is excellent agreement between the two. Of course α_C may not be 0.75, and there may be a contribution due to the effect of iR_u drop which was not accounted for in the experimental procedure. The effect of iR_u drop can be represented by the equation:

$$E_{peak} = E - C_1 \sqrt{v} \qquad \{4.31\}$$

where C_1 (V) is a constant. If 4.31 is combined with 4.30 and the values for R, T, F, n', and α_C (assumed to be 0.5) are inserted then an expression containing both the kinetic and IR drop terms is obtained,

$$E_{peak} = C_2 - 0.026\ln(19.14v) - C_3\sqrt{v} \qquad \{4.32\}$$

where C_2 (V) and C_3 (V) are both constants. Figure 4.16 shows the result of a fit with E_{peak} with C_2 = -250.0 mV and C_3 = 3.0 mV. Again the fit is in very good agreement with the experimental data.

From the cyclic voltammogram for copper at v = 25 mV s^{-1} a current of 14.30 µA is obtained for the reduction of Cu^+. This is related to the diffusion co-efficient by equation 4.25. This gives a value of D_{CV} = 3.63 × 10^{-6} cm^2 s^{-1} assuming α_C is equal to 0.5.

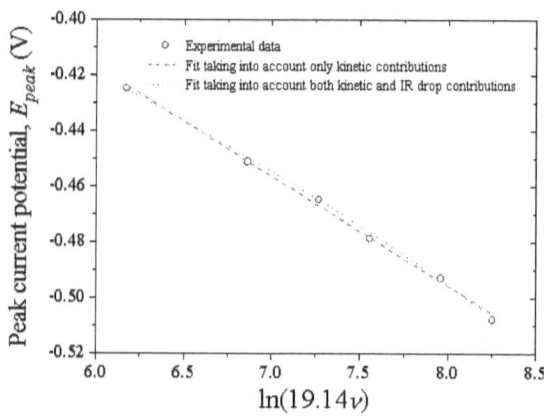

Figure 4.16 The variation of E_{peak} versus ln(19.14v) for copper. The area of the electrode was 0.06 cm^2 and the solution comprised 5.40 mM $Cu(NO_3)_2H_2O$ and 500.0 mM of $Ba(NO_3)_2$ dissolved in DMSO. The solution was maintained at 30 °C during the experiment.

A potential step experiment (chronoamperometry) was also performed with the copper solution. The potential was held at 0.0 V for 10 s and then stepped to -1.0 V for 10 s. Figure 4.17 shows the variation of current with time for this experiment. According to the *Cottrell* equation (4.27) a straight line should be obtained if the current is plotted against $t^{-1/2}$ and the gradient, Δ, of the line is related to D_{SV} by equation 4.26. The results of this transformation are shown in figure 4.18 where the linear regression line fit yielded a value of Δ of 49.3 $\mu A\ s^{-1/2}$. By applying equation 4.26, D_{SV} was found to be $7.97 \times 10^{-6}\ cm^2\ s^{-1}$, significantly different from the value obtained from cyclic voltammetry.

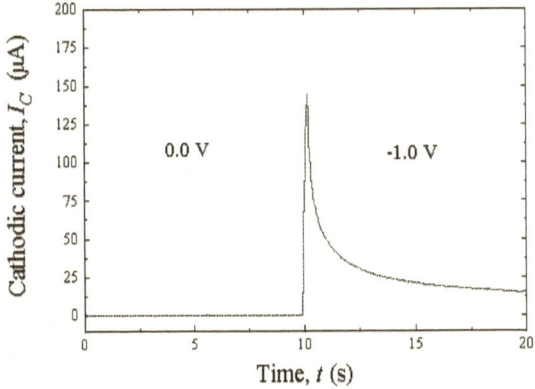

Figure 4.17 Step voltammogram of copper in DMSO. The area of the electrode was 0.06 cm^2 and the solution comprised 5.40 mM $Cu(NO_3)_2H_2O$ and 500.0 mM of $Ba(NO_3)_2$ dissolved in DMSO. The solution was maintained at 30 °C during the experiment. The potential was held at 0.0 V (versus Ag psuedo reference) for 10 s and then stepped to -1.0 V for 10 s.

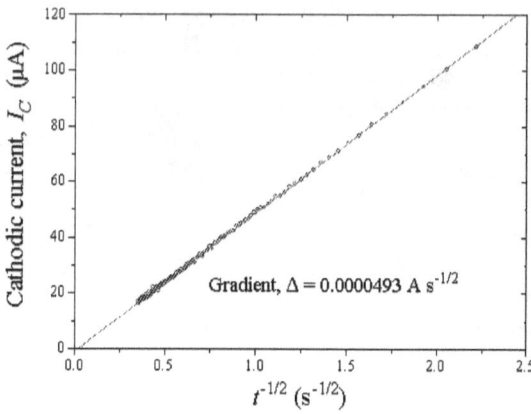

Figure 4.18 The cathodic current versus the inverse root of the deposition time for copper. The gradient is proportional to the diffusion constant. The area of the electrode was 0.06 cm² and the solution comprised 5.40 mM Cu(NO₃)₂H₂O and 500.0 mM of Ba(NO₃)₂ dissolved in DMSO. The solution was maintained at 30 °C during the experiment.

From the results presented above it is clear that the electrochemistry of copper in DMSO is not a reversible process - the separation between the reduction and oxidation peaks is too large and E_{peak} is dependent on v. Figure 4.19 shows the relationship between I_{peak} and $v^{1/2}$ for the reaction $Cu^+/Cu^{(0)}$. If the electrochemistry of copper in DMSO was in fact reversible or irreversible then this relationship would be a linear one. From the figure it is clear that at low sweep rates the relationship seems to be linear, but at higher sweep rates the linearity does not persist. This profile is typical of a *quasi-reversible* electrochemical process.

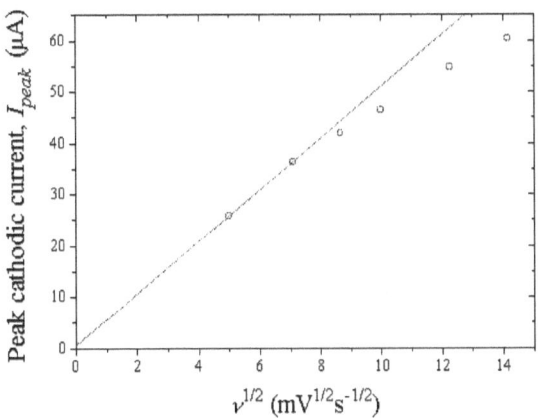

Figure 4.19 The peak cathodic current versus the $v^{1/2}$ for copper. The area of the electrode was 0.06 cm^2 and the solution comprised 5.40 mM Cu(NO$_3$)$_2$H$_2$O and 500.0 mM of Ba(NO$_3$)$_2$ dissolved in DMSO. The solution was maintained at 30 °C during the experiment.

4.3.3 Electrochemistry of Lead

The cyclic voltammograms of lead in DMSO (figure 4.20) are similar to those of copper in DMSO in that the separation of the reduction and oxidation peaks was found to be greater than 56.6/n mV. In fact the separation was found to be ~ 104 mV. The electrochemistry of Pb^{2+}, however, is different from Cu^{2+} in that the reduction of the lead cations to lead metal (Pb$^{(0)}$) occurred in a single stage (Pb^{2+} + 2e$^-$ → Pb$^{(0)}$) indicating that Pb$^+$ ions are not stable. It can be seen from the figure that the redox potential of Pb^{2+} occurs at -0.55 V. Again, a stripping peak was observed with a peak current potential of ~ 0.18 V.

The observed negative shift in E_{peak} with increasing v can again be explained using the theory of an irreversible system. As with the copper two fits were applied to the shift. The first which took into account only the effects of reaction kinetics yielded the following relationship,

$$E_{peak} = -0.475 - 0.015\ln(38.92v) \qquad \{4.33\}$$

Theoretically the gradient of the fit should be -0.013 V (assuming α_C to be 0.5) which is in good agreement with the experimental value of -0.015 V. When the theoretical value was used and equation 4.34 fitted to the data the values of C_4 and C_5 were calculated to be -491.0 mV and 0.38 mV. Figure 4.21 shows the experimental data and the curves resulting from the two fits - both the fits yielding good approximations to the data.

$$E_{peak} = C_4 - 0.0131\ln(38.92v) - C_5\sqrt{v} \quad \{4.34\}$$

The peak cathodic current at a sweep rate of 25 mV s^{-1} was determined to be 28.74 µA for the reduction of Pb^{2+} which when converted into a diffusion rate (using equation 4.25) a value of $D_{CV} = 1.05 \times 10^{-6}$ cm^2 s^{-1} was obtained.

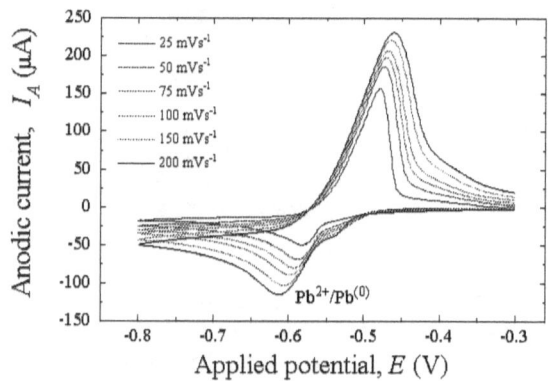

Figure 4.20 *The variation of cathodic current versus applied potential for different sweep rates for lead. The area of the electrode was 0.06 cm^2 and the solution comprised 5.04 mM Pb(NO$_3$)$_2$ and 500.0 mM of Ba(NO$_3$)$_2$ dissolved in DMSO. The solution was maintained at 30 °C during the experiment.*

A chronoamperometry experiment was performed whereby the applied potential was held at -0.4 V for 10 s and then stepped to -1.0 V for 10 s. The results of the experiment are depicted in figure 4.22 where the cathodic current has been plotted against the inverse square root of the deposition time. The gradient of the line was calculated to be 51.3 µA s$^{-1/2}$ which yielded a value for the diffusion rate, D_{SV}, of 2.48×10^{-6} cm^2 s^{-1}.

The data collected concerning the electrochemistry of lead in DMSO show that the electrochemical kinetics are not consistent with a reversible system. The wide separation between the reduction and oxidation peaks, and the shifting of E_{peak} with v indicate that the process is quasi-reversible. This evidence was strengthened when the peak cathodic current, I_{peak} was plotted against $v^{1/2}$ (figure 4.23). The relationship shows a inclination away from the expected relationship for a reversible system, probably reflecting the transition between a totally reversible and a totally irreversible process.

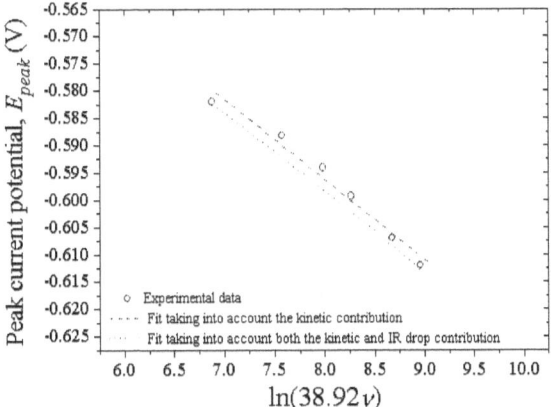

Figure 4.21 The variation of E_{peak} versus ln(38.92v) for lead. The area of the electrode was 0.06 cm^2 and the solution comprised 5.04 mM Pb(NO$_3$)$_2$ and 500.0 mM of Ba(NO$_3$)$_2$ dissolved in DMSO. The solution was maintained at 30 °C during the experiment.

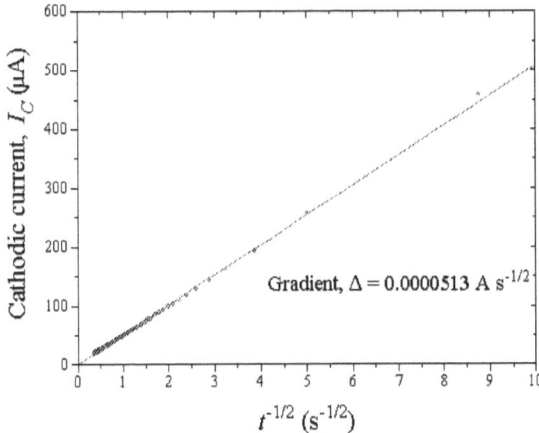

Figure 4.22 The cathodic current versus the inverse root of the deposition time for lead. The area of the electrode was 0.06 cm^2 and the solution comprised 5.04 mM Pb(NO$_3$)$_2$ and 500.0 mM of Ba(NO$_3$)$_2$ dissolved in DMSO. The solution was maintained at 30 °C during the experiment.

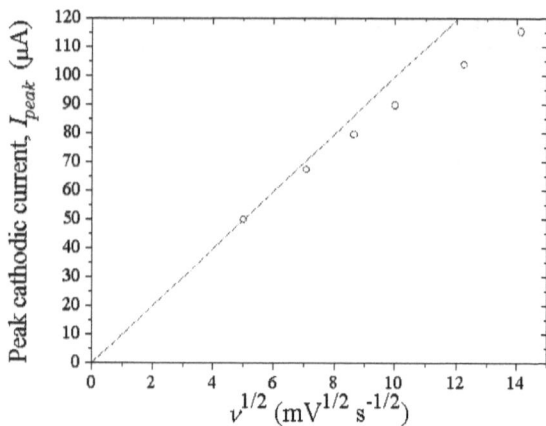

Figure 4.23 The peak cathodic current versus the root of the sweep rate for lead. The area of the electrode was 0.06 cm² and the solution comprised 5.04 mM Pb(NO₃)₂ and 500.0 mM of Ba(NO₃)₂ dissolved in DMSO. The solution was maintained at 30 °C during the experiment.

4.3.4 Electrochemistry of Thallium

Thus far, the presented analysis has determined the electrochemistry of copper and lead in DMSO. The copper displayed a two-stage process from Cu(II) to Cu(I) to copper metal, and the lead displayed a single-stage process from Pb(II) to lead metal. From cyclic voltammograms each reduction peak had its corresponding oxidation peak. The cyclic voltammogram for thallium in DMSO is quite different (figure 4.24). At lower sweep rates there were two reduction peaks observed, occurring closely together with respective redox potentials of -0.75 V and -0.80 V. At higher sweep rates the latter of the two peaks was 'drowned out' by the prior. Similarly, there were two oxidation peaks present, also very close together, with one more dominant at lower sweep rates than the other. These artefacts may be explained by the following.

A prerequisite for the growth of thallium on the silver substrate is the formation of thermodynamically stable nuclei on the surface. The formation of such nuclei requires a potential more negative than that required to reduce Tl^+ cations, and leads to what is known as a *nucleation overpotential*, i.e. the potential has to be more negative to deposit Tl on Ag than on Tl itself. Therefore, the reduction peak with an observed redox potential of -0.80 V is due to the creation of nucleation sites on the Ag substrate. Once nucleation sites have been established then deposition of Tl upon Tl may proceed, requiring a potential to reduce Tl^+ cations (-0.75 V). The peak oxidation current potentials of ~ -0.70 V and ~ -0.63 V (at 100 mV s^{-1}) correspond to the stripping of the Tl from a Ag layer, and the Tl from a Tl layer, respectively.

As before, the shift in the E_{peak} with v was fitted using a kinetic based equation and an equation including both kinetic and IR drop effects. The fit based on equation 4.30 (kinetic contributions only) yielded $E_{peak} = -0.631 - 0.021\ln(19.12v)$. The gradient of -0.021 (V) is low when compared to the theoretical value of -0.026 (V) (assuming α_C is 0.5) but is acceptable when considering that α_C lies in the range 0.25 → 0.75.

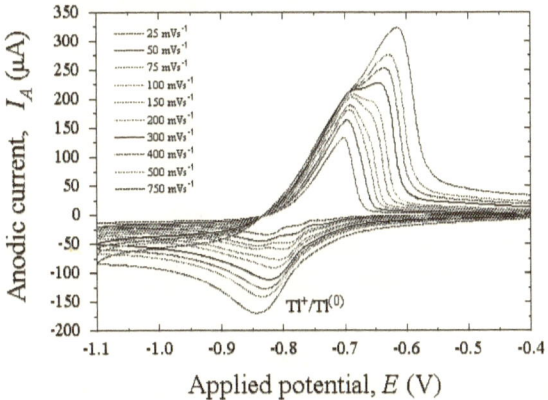

Figure 4.24 The variation of cathodic current versus applied potential for different sweep rates for thallium. The area of the electrode was 0.06 cm^2 and the solution comprised 5.03 mM TlNO$_3$ and 500.0 mM of Ba(NO$_3$)$_2$ dissolved in DMSO. The solution was maintained at 30 °C during the experiment.

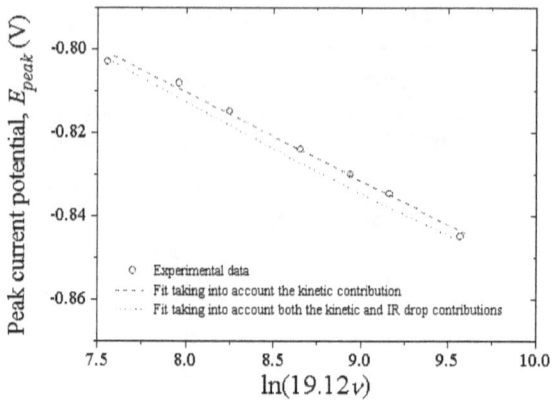

Figure 4.25 The variation of E_{peak} versus $\ln(19.14v)$ for thallium. The area of the electrode was 0.06 cm^2 and the solution comprised 5.03 mM TINO$_3$ and 500.0 mM of Ba(NO$_3$)$_2$ dissolved in DMSO. The solution was maintained at 30 °C during the experiment.

An alternative fit was also applied in which both the kinetic and iR_u drop effects were taken into account. The fit (equation 4.32) produced values for C_2 and C_3 of -611 mV and -0.5 mV. Figure 4.25 shows how both the fits measure up against the experimental data. Again, both approaches give good approximations to the data.

The peak cathodic current at a sweep rate of 25 mV s^{-1} was determined to be 10.68 µA for the reduction of Tl$^+$ which when converted into a diffusion rate (using equation 4.25) a value of $D_{CV} = 8.27 \times 10^{-6}$ cm^2 s^{-1} was obtained.

A chronoamperometry experiment was performed whereby the applied potential was held at -0.5 V for 10 s and then stepped to -1.2 V for 10 s. The results of the experiment are depicted in figure 4.26 where the cathodic current has been plotted against the inverse square root of the deposition time. The gradient of the line was calculated to be 56.1 µA s$^{-1/2}$ which yielded a value for the diffusion rate, D_{SV}, of 8.40×10^{-6} cm^2 s^{-1}.

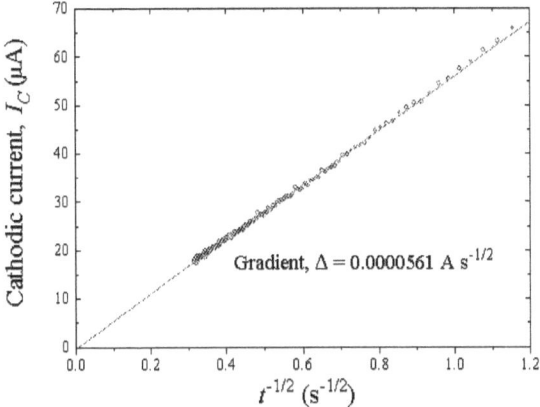

Figure 4.26 The cathodic current versus the inverse root of the deposition time for thallium. The area of the electrode was 0.06 cm^2 and the solution comprised 5.03 mM TINO$_3$ and 500.0 mM of Ba(NO$_3$)$_2$ dissolved in DMSO. The solution was maintained at 30 °C during the experiment.

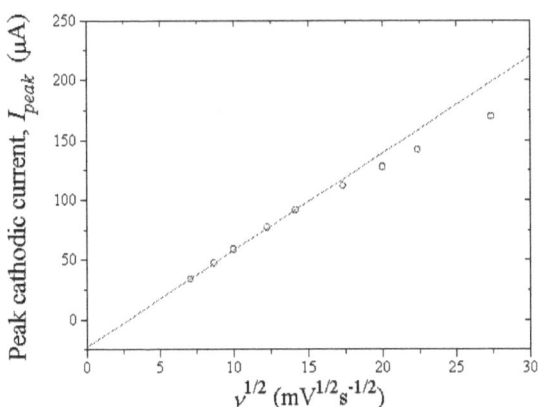

Figure 4.27 The peak cathodic current versus the root of the sweep rate for thallium. The area of the electrode was 0.06 cm^2 and the solution comprised 5.03 mM TINO$_3$ and 500.0 mM of Ba(NO$_3$)$_2$ dissolved in DMSO. The solution was maintained at 30 °C during the experiment.

4.3.5 Electrochemistry of Bismuth

The cyclic voltammograms for Bi^{3+} in DMSO (figure 4.28) show, better than the other elements investigated, the profile expected for experiments during which metal deposition occurs. The important feature of metal deposition is identified as the leading edge of the reduction peak is slightly steeper than that for a process involving only solution soluble species. This feature was clearly observed for the reduction of bismuth(III), i.e. $Bi^{3+} + 3e^- \rightarrow Bi^{(0)}$, at a sweep rate of 25 mV s^{-1}. As the sweep rate was increased the width of the reduction peaks increased due to the kinetic limitations of the process. The redox potential of $Bi^{3+}/Bi^{(0)}$ was observed to be -0.24 V, and the peak oxidation current was found to occur at ~ -0.06 V.

As with copper, lead, and thallium the shift in E_{peak} with increasing v can be modelled well with the two approaches discussed above. The first fit, taking into account the kinetic regime of an irreversible reaction only resulted in the fit $E_{peak} = 0.136 - 0.057\ln(57.42v)$ (based on 4.30). Theoretically, the gradient of the line fit should be -0.056 V (with α_C = 0.5) which is in excellent agreement with the experimentally observed value of -0.057 V.

The fit taking into account the effects of both kinetics and iR_U drop yielded $E_{peak} = 0.128 - 0.056\ln(57.42v) - 0.00035\sqrt{v}$.

The accuracy of both the fits can be seen in figure 4.29 - either approach gives an excellent agreement with experimentally observed data. The agreement with the experimental data and the separation between the peak reduction and peak oxidation currents (~ 210 mV at 25 mV s^{-1}) indicate that the electrochemistry of bismuth(III) in DMSO is irreversible.

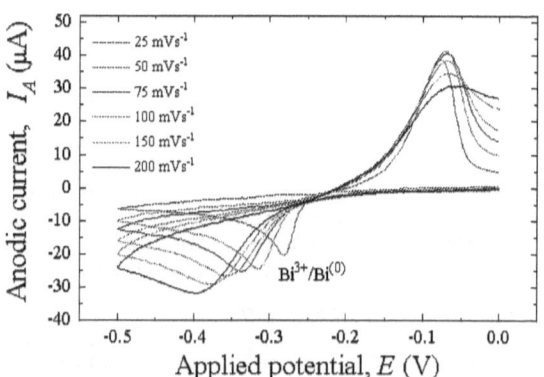

Figure 4.28 The variation of cathodic current versus applied potential for different sweep rates for bismuth. The area of the electrode was 0.06 cm^2 and the solution comprised 4.38 mM Bi(NO$_3$)$_3$ and 500.0 mM of Ba(NO$_3$)$_2$ dissolved in DMSO. The solution was maintained at 30 °C during the experiment.

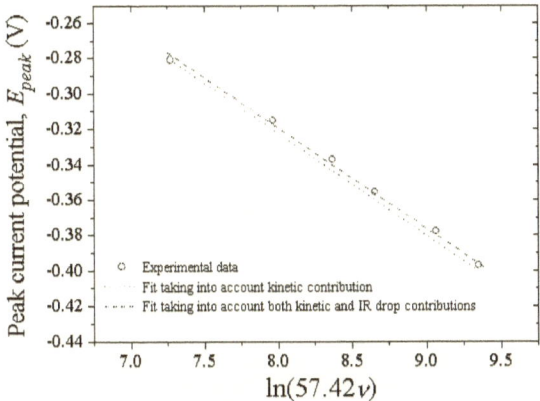

Figure 4.29 The variation of E_{peak} versus ln(57.42v) for bismuth. The area of the electrode was 0.06 cm² and the solution comprised 4.38 mM Bi(NO$_3$)$_3$ and 500.0 mM of Ba(NO$_3$)$_2$ dissolved in DMSO. The solution was maintained at 30 °C during the experiment.

The peak reduction current at 100 mV s^{-1} was determined to be 18.10 µA which resulted in a diffusion rate, D_{CV}, of 1.64×10^{-7} cm² s^{-1}. The value obtained from a chronoamperometry experiment performed (D_{SV}), where the applied potential was held at -0.1 V for 10 s and then stepped to -1.0 V for 10 s, was 1.70×10^{-7} cm² s^{-1} which is in agreement with the value obtained from cyclic voltammetry. Figure 4.30 shows the cathodic current versus the inverse square root of the deposition time obtained for the -1.0 V step of the chronoamperometry experiment performed.

Figure 4.31 shows the peak cathodic current versus the square root of the sweep rate. As with copper, lead, and thallium the plot indicates a transition from reversible to irreversible behaviour characteristic of a quasi-reversible process. This conclusion is consistent with the data presented above for the electrochemistry of bismuth(III) in DMSO.

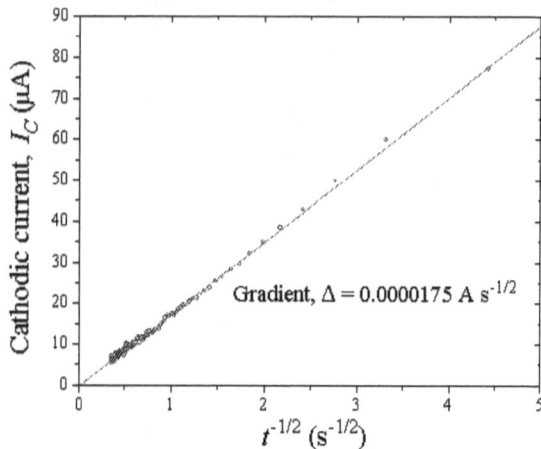

Figure 4.30 The cathodic current versus the inverse root of the deposition time for bismuth. The area of the electrode was 0.06 cm² and the solution comprised 4.38 mM Bi(NO₃)₃ and 500.0 mM of Ba(NO₃)₂ dissolved in DMSO. The solution was maintained at 30 °C during the experiment.

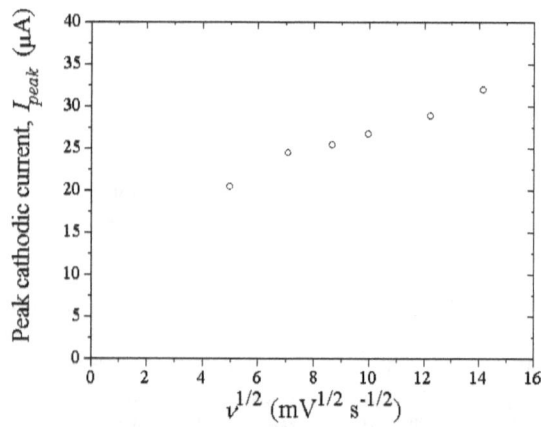

Figure 4.31 The peak cathodic current versus the root of the sweep rate for bismuth. The area of the electrode was 0.06 cm² and the solution comprised 4.38 mM Bi(NO₃)₃ and 500.0 mM of Ba(NO₃)₂ dissolved in DMSO. The solution was maintained at 30 °C during the experiment.

4.3.6 Electrochemistry of Mercury

Figure 4.32 shows cyclic voltammograms obtained at a range of sweep rates for 4.93 mM HgNO₃ dissolved in DMSO. Firstly, clearly the voltammograms did not extend far enough to include the stripping peak of the Hg metal ($Hg^{(0)}/Hg^+$). This was because if the potential was taken too high then oxidation of the Ag electrode would have occurred. Tables that relate the redox potentials of the elements to the SHE indicate that the redox potentials for mercury(I) and silver(I) are very close together. From figure 4.32 it can be seen that the redox potential for $Hg^+ + e^- \rightarrow Hg^{(0)}$ in DMSO is ~ 0.00 V (the redox potential for Ag^+ in this system was observed to occur at ~ +0.02 V).

As observed above, the shift in E_{peak} with increasing v was modelled well with two different approaches discussed previously. The first fit, taking into account the kinetic contributions exhibited by irreversible processes resulted in the fit $E_{peak} = 0.025 - 0.010\ln(19.12v)$. The gradient of -0.010 V does not agree well with the expected value of -0.026 V (assuming α_C = 0.5). However, when a fit taking into account both kinetic and iR_U drop contributions was performed a good fit of the experimental data was achieved. The resulting relationship between E_{peak} and v was determined to be

$$E_{peak} = 0.096 - 0.026\ln(19.12v) + 0.0013\sqrt{v}.$$

Figure 4.33 shows how the two different methods compare to the experimental data.

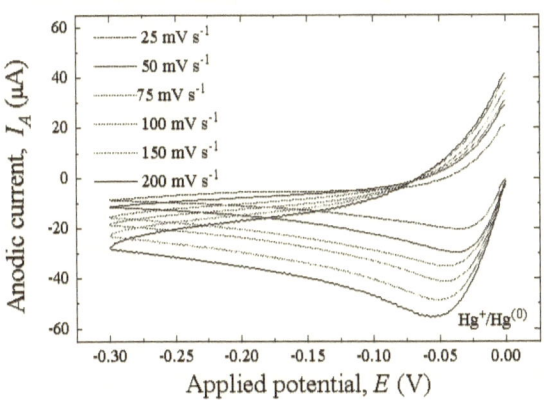

Figure 4.32 The variation of cathodic current versus applied potential for different sweep rates for mercury. The area of the electrode was 0.06 cm² and the solution comprised 5.4 mM HgNO₃ and 500.0 mM of Ba(NO₃)₂ dissolved in DMSO. The solution was maintained at 30 °C during the experiment.

The peak reduction current at a sweep rate of 100 mV s^{-1} was found to be 41.13 µA which resulted in a diffusion rate, D_{CV}, of 2.79 × 10^{-6} cm^2 s^{-1}. The value obtained from a chronoamperometry experiment performed (D_{SV}), where the applied potential was held at 0.0 V for 10 s and then stepped to -0.3 V for 10 s, was 3.23 × 10^{-6} cm^2 s^{-1}, which is in reasonable agreement with the value obtained from cyclic voltammetry. Figure 4.34 shows the cathodic current versus the inverse square root of the deposition time obtained for the -0.3 V step of the chronoamperometry experiment performed. The gradient, Δ, of 0.0000175 A s$^{-1/2}$ was related to the diffusion constant, D_{SV}, by equation 4.26.

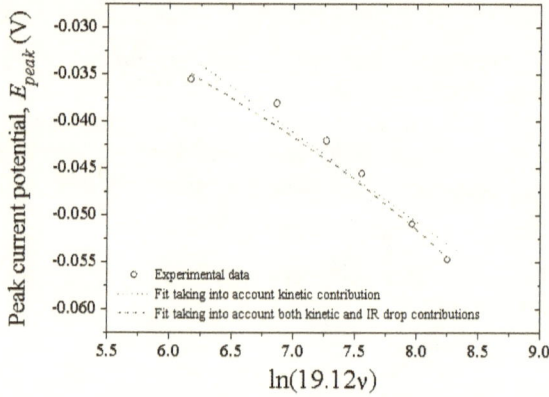

Figure 4.33 The variation of E_{peak} versus ln(19.12v) for mercury. The area of the electrode was 0.06 cm^2 and the solution comprised 4.38 mM HgNO$_3$ and 500.0 mM of Ba(NO$_3$)$_2$ dissolved in DMSO. The solution was maintained at 30 °C during the experiment.

As with the electrochemistry of copper, thallium, lead, and bismuth, the electrochemistry of mercury in DMSO is not reversible. This is reinforced by the data presented in figure 4.35 which shows the variation of I_{peak} with $v^{1/2}$ for the reaction Hg$^+$/Hg$^{(0)}$. The relationship appears linear at lower sweep rates (≤ 100 mV s^{-1}) and then diverges from linear behaviour at higher sweep rates. This is consistent of a *quasi-reversible* electrochemical process.

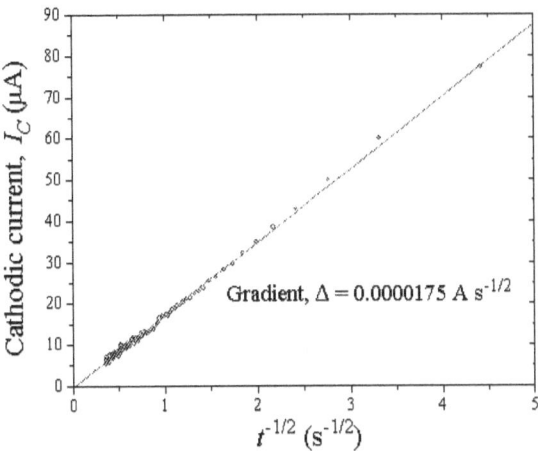

Figure 4.34 The cathodic current versus the inverse root of the deposition time for mercury. The area of the electrode was 0.06 cm² and the solution comprised 4.93 mM HgNO₃ and 500.0 mM of Ba(NO₃)₂ dissolved in DMSO. The solution was maintained at 30 °C during the experiment.

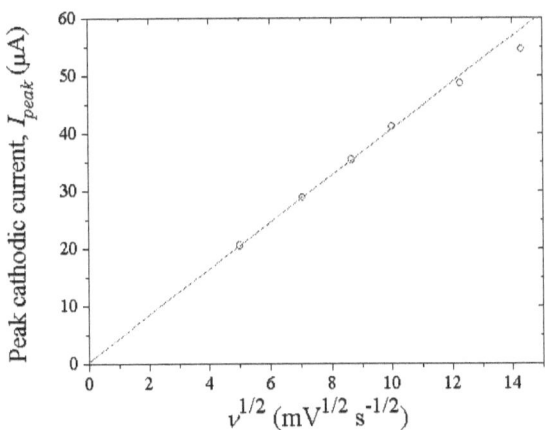

Figure 4.35 The peak cathodic current versus the root of the sweep rate for mercury. The area of the electrode was 0.06 cm² and the solution comprised 4.93 mM HgNO₃ nitrate and 500.0 mM of Ba(NO₃)₂ dissolved in DMSO. The solution was maintained at 30 °C during the experiment.

4.3.7 Electrochemistry of Barium, Strontium, and Calcium.

Since the alkaline earth metals, Ca, Sr, and, Ba, are violently attacked by all aqueous solutions, it is very difficult to obtain values of the redox potentials for the electrochemical reactions in solutions containing water [18]. Attempts to determine the redox potentials of Ca^{2+} and Sr^{2+} were unsuccessful when cyclic voltammetric experiments were performed on DMSO solutions containing the nitrate salts of calcium and strontium. This may be due to the unstability of these elements in water containing solutions, but may also be because of the quick conversion from metals to oxides, and then to carbonates. This would result in deposition of insulating layers of CaO and SrO which would quickly prevent further deposition, and prevent an observable deposition current from being measured. This is confirmed by data obtained for the electrodeposition of the superconductor constituents presented in the proceeding section.

In the following chapter cyclic voltammetry performed on a superconductor precursor solution containing nitrate salts of Tl, Pb, Sr, Ca, and Cu is presented. Because of the conductive metallic deposits, resulting from the Tl, Pb, and Cu metal, deposition of Sr and Ca was possible. Figure 4.36 shows a section of the cyclic voltammogram obtained where the peak was determined to be due to the reduction of Ca^{2+} and Sr^{2+}. From the figure there appears to be two overlying peaks indicating that the cations of calcium and strontium have very similar redox potentials (-1.80 V) which is confirmed by reference [19]. The thinness of the peak confirms that an insulating layer was indeed deposited. The determination of diffusion rates via chronoamperometry was not successful for the reasons discussed above.

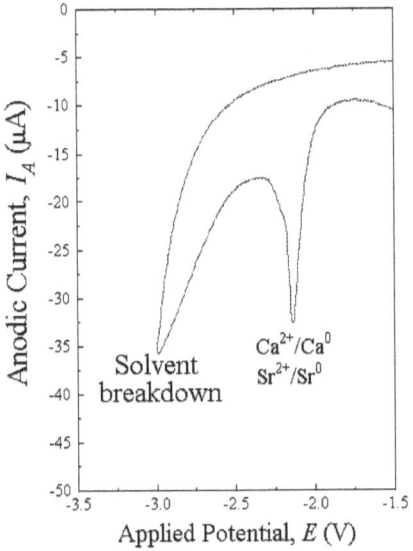

Figure 4.36 Cyclic voltammogram performed on a solution comprised from 6.5 mM TlNO₃, 2.0 mM Pb(NO₃)₂, 250.0 mM Sr(NO₃)₂, 450 mM Ca(NO₃)₂H₂O, and 12.0 mM Cu(NO₃)₂H₂O dissolved in DMSO. The experiment was performed in a dry box and the cell regulated at 30 °C. The sweep rate was 100 mV s⁻¹.

Cyclic voltammetry was successful in obtaining a peak due to the reaction $Ba^{2+} + 2e^- \rightarrow Ba^{(0)}$. This is shown in figure 4.37 indicating that the redox potential of $Ba^{2+}/Ba^{(0)}$ is -1.60 V. From the figure it can also be seen that there was no stripping peak observed. Barium metal behaves similarly to calcium and strontium metal, in as far as they all absorb oxygen rapidly from the atmosphere to form oxides. If the barium deposit quickly evolved into an insulating oxide layer rather than remaining a conductive metallic deposit then no stripping peak would be expected - an irreversible process. The source of the oxygen would have been from the dry box atmosphere with small amounts remaining in the solution after irradiation with ultrasound. Also observed from figure 4.37 was the reduction of oxygen at ~ -1.0 V which confirms that oxygen was still present in the solution.

Figure 4.38 shows the results from a chronoamperometry experiment. The gradient, Δ, of the straight line fit was found to be 9.14×10^{-5} A s$^{-1/2}$ which translates into a value of D_{SV} of 5.61×10^{-6} cm² s⁻¹ assuming $n = 2$. From cyclic voltammetry,

D_{CV} was calculated to be 1.43×10^{-6} cm^2 s^{-1} which was determined from a peak current of 21.10 µA at a sweep rate of 100 mV s^{-1}.

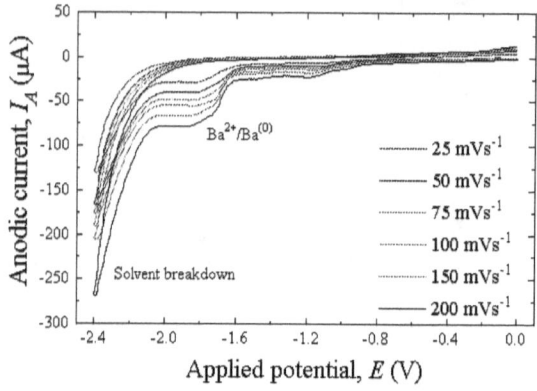

Figure 4.37 The variation of cathodic current versus applied potential for different sweep rates for barium. The area of the electrode was 0.06 cm^2 and the solution comprised 5.01 mM Cu(NO$_3$)$_2$H$_2$O and 500.0 mM of C$_{16}$H$_{36}$BF$_4$N dissolved in DMSO. The solution was maintained at 30 °C during the experiment.

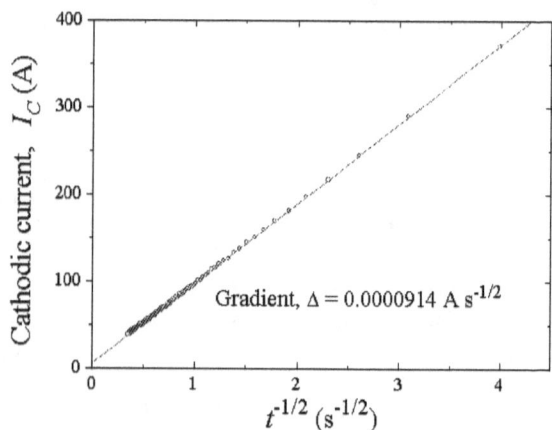

Figure 4.38 The cathodic current versus the inverse root of the deposition time for barium. The area of the electrode was 0.06 cm^2 and the solution comprised 5.01 mM Cu(NO$_3$)$_2$H$_2$O and 500.0 mM of C$_{16}$H$_{36}$BF$_4$N dissolved in DMSO. The solution was maintained at 30 °C during the experiment.

4.4 Electrodeposition of Superconductor Constituents

4.4.1 What is Electrodeposition?

The mechanism of electrodeposition involves, initially, the reduction of a cation (i.e. Tl^+, Ba^{2+}, and Cu^+) on the substrate surface, to form an adatom. This adatom migrates over the substrate surface until an energetically favourable position is found and a nucleation site is formed. Other atoms then aggregate with it to form a new phase. Unexpectedly, the formation of the first layers dictate the structure and adhesion of the final electrodeposited film.

Generally, the potential applied to the electrode determines the rate of the reactions occurring. This in turn decides the deposition rate and therefore the structure of the film [18]. The slower the deposition rate the more perfect the crystalline structure produced. Hence, a low overpotential is preferred. The different types of structure can be split into four divisions (figure 4.39). The first occurs at slow deposition rates and so almost perfect crystalline structures are formed. As the rate increases polycrystalline structures appear. At higher rates still, nodules and dendrites are formed causing irregular film development. Material may also breakaway from the deposit. At very high rates, powders are synthesised with very low adhesion causing material to fall away from the substrate surface.

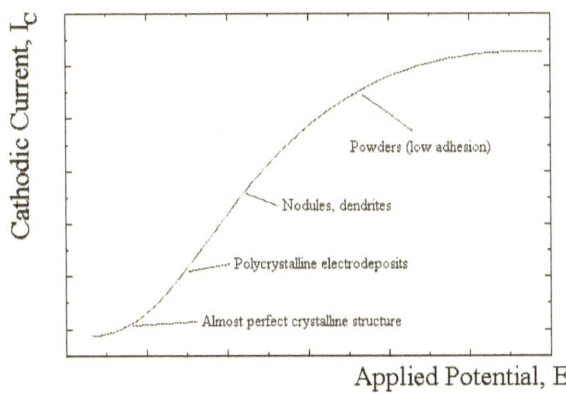

Figure 4.39 Variation of electrodeposit structure with applied potential. (adapted from reference 18)

4.4.2 Experimental

As with the cyclic and chronoamperometry experiments, extensive precautions were taken to ensure that the experimental environment was free from contamination, especially from water and oxygen. Again, a length of Ag wire was used as the reference, but a large piece of Pt foil (25.0 × 10.0 mm) covered with Pt gauze was employed as the counter electrode. Sections of polished Ag foil with one side masked with silicone rubber were used as the substrates. The schematic of the cell used, with the electrodes in place, is shown in figure 4.40. The solutions used were the same as the solutions used for the cyclic and chronoamperometry experiments except for Ba, Sr, and Ca. These solutions composed of 20 mM of the nitrate salts with 100 mM tetraethylammonium perchlorate ($C_8H_{20}ClNO_4$) as a supporting electrolyte. Each solution was ultrasonically irradiated before deposition to degas it. Deposition was caused by applying a constant potential to the substrate, with respect to the Ag reference electrode. For Cu, Pb, Tl, and Bi, the potential was held at -1.3 V for 1800 s, whereas for Ba, Sr, and Ca, the potential was held at -3.50 V for 1800 s. The morphology of the as-deposited films was analysed using an SEM.

4.4.3 Electrodeposition of Cu, Pb, Tl, Bi, Ba, Sr, and Ca

Table 4.3 shows the total charge transferred during the deposition period divided by the nitrate salt concentration, and the electrical resistivity of the different metals deposited [20]. The table shows that the amount of material deposited was observed to be loosely correlated to the electrical resistivity of the deposited metal. The exception was calcium. Calcium metal has a low electrical resistivity of 3.91 µΩ cm. Calcium metal, however, is never found in nature uncombined as it is extremely unstable. The reason that the total charge transferred was so low, therefore, was that the calcium metal had reacted with either the water or the oxygen in the solution to form a passive layer of some kind. This passive layer preventing further deposition from occurring.

The main implication from this data from the point of view of multi-metal deposition is that copper assists deposition by ensuring a reasonable film conductivity, whereas calcium hinders deposition by forming passive layers, i.e. layers with very low conductivity.

Metal	Total charge transferred per unit solution concentration (C M⁻¹)	Electrical resistivity (µΩ cm)
Copper	322	1.67
Thallium	213	18
Lead	270	21
Strontium	200	23
Barium	154	-
Bismuth	147	106.8
Calcium	7.5	3.91

Table 4.3 Total charge transferred per unit solution concentration and electrical resistivities of the superconductor constituent metals.

Figure 4.41 shows the morphologies of the as-deposited films. Calcium is not included as the amount of material deposited was so small that it was not possible to see the film using the SEM. As all the films were deposited at potentials much larger than the redox potentials of the metallic cations rough morphologies were observed due to dendritic growth. The SEM's for the bismuth and strontium films seem particularly smooth though. The smoothness of the bismuth film may be attributed to the high electrical resistivity of the metal which results in a lower deposition rate. The strontium film appears very smooth with cracking forming islands of ~ 80 µm in size. The smooth appearance may be due to the rapid oxidation of strontium metal to strontium oxide but the means by which this affects film morphology was not determined.

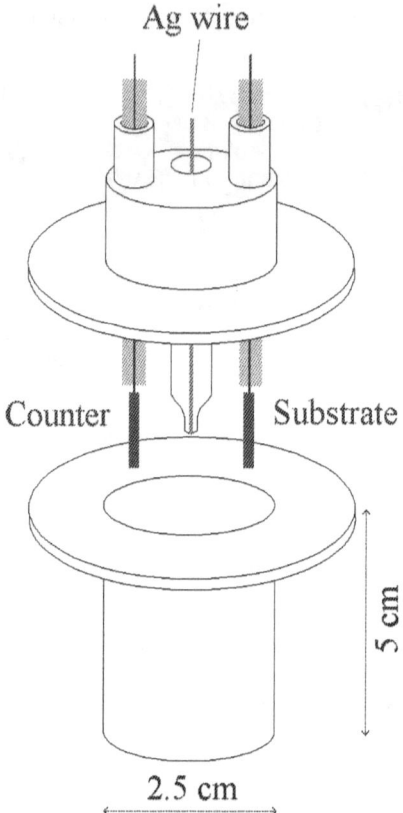

Figure 4.39 Schematic of the cell used in electrodeposition experiments.

EDS analysis of all the films showed that oxygen was present, but it was not possible to evaluate how much oxygen was included or whether the source of the oxygen was the deposition solution or the laboratory atmosphere.

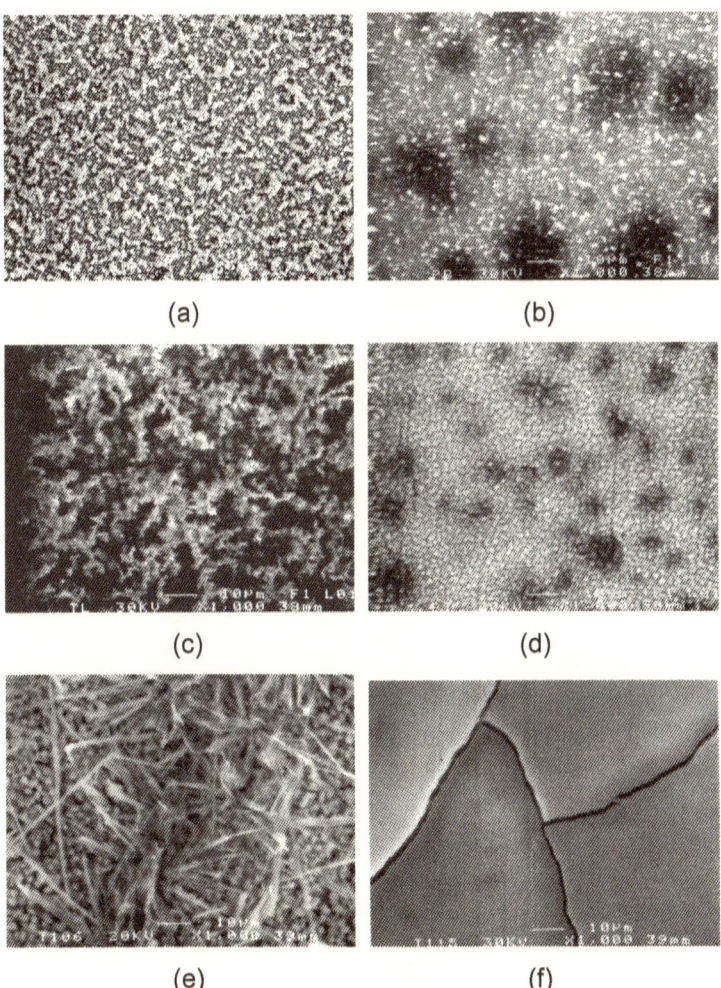

Figure 4.41 SEM micrographs of (a) copper, (b) lead, (c) thallium, (d) bismuth, (e) barium, and (f) strontium films electrodeposited from DMSO. The films were deposited onto polished Ag substrates from a fixed geometry cell regulated at 30 °C. A constant potential of -1.3 V was held for 3600 s to cause deposition of the copper, lead, thallium, and bismuth films. Whereas, a constant potential of -3.5 V was held for 1800 s to cause deposition of the barium and strontium films.

4.5 Conclusions

This chapter has described the basic theory of electrochemical processes and discussed the influencing factors upon the deposition procedure, including solution water/oxygen content, reference electrode, substrate preparation, solvent selection, and deposition rate. These factors were then applied to the experimental analysis of the basic superconductor precursor components. The results of the investigation are presented in table 4.4.

Element	Redox Potential E_{red} (V)	Diffusion Coefficient, D_{CV} (cm^2 s^{-1})	Diffusion Coefficient D_{SV} (cm^2 s^{-1})	Process type
Barium	$Ba^{2+}/Ba^{(0)}$: -1.59	1.43×10^{-6}	5.61×10^{-6}	irreversible
Bismuth	$Bi^{3+}/Bi^{(0)}$: -0.24	1.64×10^{-6}	1.70×10^{-6}	quasi-reversible
Calcium	$Ca^{2+}/Ca^{(0)}$: ~ -2.19	-	-	irreversible
Copper	Cu^{2+}/Cu^{+}: -0.08 $Cu^{+}/Cu^{(0)}$: -0.38	3.63×10^{-6}	7.97×10^{-6}	quasi-reversible
Mercury	$Hg^{+}/Hg^{(0)}$: -0.01	2.79×10^{-6}	3.23×10^{-6}	quasi-reversible
Lead	$Pb^{2+}/Pb^{(0)}$: -0.55	1.05×10^{-6}	2.48×10^{-6}	quasi-reversible
Strontium	$Sr^{2+}/Sr^{(0)}$: ~ -2.19	-	-	irreversible
Thallium	$Tl^{+}/Tl^{(0)}$: -0.75	8.27×10^{-6}	8.40×10^{-6}	quasi-reversible

Table 4.4 Fundamental electrochemical parameters for superconductor precursor constituents.

The electrochemical data for copper, lead, thallium, bismuth, and mercury were easily explained by applying basic results of the theory for irreversible systems. A term derived from the effects of IR drop was also included. Other information concluded that they were actually quasi-reversible systems. The limited data obtained concerning the electrochemistry of strontium, barium, and calcium suggested that they were all irreversible systems. The instability of the basic metals though may indicate that the process could be split into an electrochemical step and a chemical step, a process known as an EC mechanism.

The large range of redox potentials has consequences for multi-metal co-deposition. It is known that the larger the applied potential, compared to the redox potential, the more dendritic (porous) the film produced will be. Therefore, when applying large potentials (~ - 3 V) in order to deposit metals of calcium, strontium, and barium, the deposition of other constituents in the solution occurs so fast that dendritic growth is inevitable. This problem is observed and discussed in subsequent chapters.

From the electrodeposition experiments performed it was clear that copper and calcium metal would play major roles in determining film morphology and stoichiometry of multi-metal films. Copper has the lowest electrical resistivity of the superconductor constituents and therefore is easily deposited and maintains the conductivity of the growing film assisting further deposition. The role of calcium is exactly the opposite. Calcium is difficult to deposit and, once deposited, reacts quickly to form a passive layer, hindering further deposition. These two extreme effects of copper and calcium upon film development work against each other making control of film stoichiometry a non-trivial task.

References

1 A. F. Silva ed., *Trends in Interfacial Electrochemistry*, Proceedings of NATO ASI (1984), Reidel, Dordrecht, 1985.

2 S. R. Morrison, Electrochemistry at Semiconductor and Oxidised Metal Electrodes, Plenum, New York, 1980.

3 K. Uosaki and H. Kita, *Modern Aspects of Electrochemistry*, Plenum, New York, Vol. 18, 1986, ed. R. E. White, J. O. 'M. Bockris, and B. E. Conway, pp. 1-60.

4 A. Hamnett, in *Comprehensive Chemical Kinetics*, ed. R. G. Compton, Elsevier, Amsterdam, Vol. 27, 1987, Chapter 2.

5 D. C. Grahame, *Ann. Rev. Phys. Chem.*, 6 (1955) 337.

6 R. Parsons, *Modern Aspects of Chemistry*, Butterworths, London, Vol. 1, 1954, ed. J. O. 'M. Bockris and B. E. Conway, pp. 103-179.

7   R. Parsons, *Advances in Electrochemistry and Electrochemical Engineering*, ed. P. Delahay and C. W. Tobias, Wiley, New York, Vol. 1, 1961, pp. 1-64.

8   S. Trasatti, *Modern Aspects of Electrochemistry*, Plenum, New York, Vol. 13, 1979, ed. B. E. Conway and J. O. 'M. Bockris, pp. 81-206.

9   H. L. F. von Helmholtz, *Ann. Physik*, 89 (1853) 211; 7 (1879) 337.

10 G. Gouy, *Compt. Rend.*, 149 (1910) 654.

11 D. L. Chapman, *Phil. Mag.*, 25 (1913) 475.

12 O. Stern, *Z. Elektrochem.*, 30 (1924) 508.

13 D. C. Grahame, *Chem. Rev.*, 41 (1947) 441.

14 J. O. 'M. Bockris, M. A. Devanathan, and K. Müller, *Proc. R. Soc.*, A274 (1963) 55.

15 G.Goodlett, private communication.

16 Aldrich, Catalogue Handbook of Fine Chemicals, 1994.

17 Adapted from *CRC Handbook of Chemistry and Physics*, 70th Edition, 1989, page E38.

18 C. M. A. Brett and A. M. O. Brett, *Electrochemistry: Principles, Methods, and Applications*, Oxford University Press, 1994, pp. 341-343.

19 CRC Handbook of Chemistry and Physics, 70th Edition, 1989, page D151.

20 CRC Handbook of Chemistry and Physics, 70th Edition, 1989, page F146.

Chapter 5:

Experimental Considerations for the Electrodeposition of Superconductor Precursor Films

5 Experimental Considerations for the Electrodeposition of Superconducting Precursor Films

5.1 Introduction

One of the most demanding challenges to film technology is the deposition of superconducting materials. The complexity of the processes involved is paralleled by the complex structure of the compounds formed. Ideally, any technique employed would be versatile, and relatively cheap as well as reliable. Electrodeposition, which has all of these attributes, also has another major attraction: electrochemical deposition techniques can be readily implemented as part of a continuous process. Electrodeposition also offers:

- a low cost, low energy technique;
- deposition onto complex (non-planar) shapes;
- the freedom from carbon-based impurities that could contaminate grain boundaries.

Since the discovery of high temperature superconductors in 1986 by Wu *et al.* [1] research into their fundamental properties has been well established. Now that a large number of superconducting materials has been studied and their production methods determined the time is appropriate for a more concentrated effort towards the application of these materials. Many useful applications have already been proposed such as: low-loss power cables, lossless energy storage systems, superfast electronics, superfast transmission lines, as well as exotic projects like magnetically levitated transport systems, and magnetic confinement of the plasma created in fusion power reactors. For these proposals to be realised large-scale manufacturing techniques for the production of superconducting materials must be developed.

In this chapter techniques applied thus far to the electrodeposition of HTSC precursor films are reviewed. The experimental parameters and their effect upon the electrodeposition of high temperature superconductor precursor materials are discussed with particular reference made to the film stoichiometry. The effects are illustrated by discussing their influence upon the deposition of a variety of precursor films, e.g. Bi-Sr-Ca-Cu, Ba-Ca-Cu, and Tl-Ba-Ca-Cu, and Tl-Pb-Sr-Ca-Cu. Moreover, the removal of their influence upon the procedure is discussed. These lessons are then applied to the deposition of Bi-Sr-Ca-Cu films with the desired 2212 stoichiometry. The deposition of layers containing only one element is relatively straightforward if compared against the difficulties in fabricating layers containing multiple elements with particularly desired relative abundances.

5.2 Electrodeposition of High Temperature Superconductors

The first reports towards with the electrochemical synthesis of high temperature superconducting films were published during the late 1980's. In 1988 Zurawski *et al*. [2], succeeded in producing Ba-Cu-O films on Cu substrates. The technique employed, which was very different to all subsequent reports, was cyclic voltammetry. The voltage applied to the substrate was cycled between -1.4 V and +0.725 V (against a Ag/AgCl reference electrode) at a rate of 50 mVs^{-1}. This was performed in a barium hydroxide ($Ba(OH)_2$) solution, and thus, as the Cu was oxidised and reduced, Ba was also reduced and therefore incorporated into the substrate. Zurawski *et al*. had intended to go further and add yttrium to the bath to yield Y-Ba-Cu-O films but, as yet, no follow-up reports have been published by them.

In 1989 the first report of the electrodeposition of superconducting films was published. Maxfield *et al*. [3] succeeded in depositing 1-2 µm $Bi_2Sr_2CaCu_2O_8$ and $(Pb,Bi)_2Sr_2CaCu_2O_8$ films on Ag coated MgO substrates. This was achieved by deposition at a constant applied potential of -4.0 V, against a Ag/Ag$^+$ reference, from a solution of nitrate salts of the constituent metals dissolved in dimethylsulfoxide (DMSO). The researchers also discussed the dehydration of these salts, unlike any other group, with the use of P_2O_5. The as-deposited films were dendritic, ductile, highly conductive, and nearly uniform in composition across the film with no more than a 6 % variation in relative metal abundances. These films also displayed a very high void space of approximately 60 %. The films were then heated in a flowing O_2/N_2 (1:1) atmosphere at 850 - 875 °C for 2-5 minutes and cooled to 450 °C in 2 hours to form dense, and granular, layered copper oxide superconductors. Resistivity measurements showed that both types of film (those with Pb and those without) had a T_C of 85 K. Also transport measurements indicated critical current densities of 300 - 400 A cm^{-2} at 40 K in zero field. Ultimately, the paper provided evidence that an electrochemical route to polycrystalline high temperature superconductors is indeed feasible.

Since the above publication several other groups around the world have attempted electrochemical synthesis of superconductors with varying degrees of success. Pawar et al. [4] deposited Bi-Sr-Ca-Cu alloyed films from an acetone solution, containing the nitrate salts, onto fluorine doped tin oxide (FTO) coated MgO substrates by applying a potential of -1.35 V to -1.55 V (versus a standard calomel electrode). The effect of complexing agents such a sodium citrate, ethylenediammine tetraacetic acid (EDTA), tartaric acid and sodium nitrate upon the deposition process was also investigated. The 4-6 μm thick as-deposited films were dense, well-adhered, and dark in colour. After heating at 750 °C in air for 1 hour the oxide films had a very high T_C of 95 K. Sodium citrate and EDTA were found to be suitable complexing agents. Pawar et al. have also investigated the electrodeposition of Dy-Ba-Cu alloyed films from an aqueous solution (doubly distilled water) onto steel and FTO coated MgO [5]. Because of interdiffusion, during heating, with the steel the study of these films was terminated. The films deposited on MgO, however, were heated at 680 °C in air which yielded a T_C of 100 K.

Weston et al. have reported studies into the electrodeposition of Y-Ba-Cu-O and Er-Ba-Cu-O thin films [6,7]. Research described in reference [6] explains how Y-Ba-Cu-O films were deposited from a solution of the relevant nitrate salts dissolved in dimethyl formamide (DMF). Deposition was performed at a constant potential of -5.0 V (versus Ag/AgCl) onto silver coated copper foils using a flat-cell. A T_C of 91 K in multi-phase films was achieved after heating at 920 °C in flowing O_2 for two hours, followed by a low temperature anneal to improve inter-grain connectivity as well as optimisation of oxygen stoichiometry. Following this work the group deposited films of Y-Ba-Cu-O and Er-Ba-Cu-O using a pulsed potential technique. Again, the solution was a mixture of the nitrate salts, dissolved in DMF, but the applied potential was pulsed between 0.0 V to -17.0 V at a rate of 10 - 20 Hz. Another departure from other approaches was that a two-electrode cell was used (the counter electrode filling the role of a reference also), whereas previously all investigations were reportedly carried out using a standard three-electrode set-up. The problems with this set-up are that the interfacial potential is poorly controlled resulting in poor reproducibility of the experimental procedure. Yttrium-based superconducting films had a T_C of 93 K, whereas the Erbium-based superconducting films had a T_C of 80 K. Both were deposited on zirconia and showed significant c-axis alignment perpendicular to the substrate. However, the authors did point out the key role that the concentration of Cu plays in providing the electrical conductivity necessary to allow co-deposition of Y/Er, and Ba. Also, noted was the difficulty in integrating stirring with the deposition process. Owing to the wide disparity in the reduction potentials of Cu(II) with respect to Ba(II) and Y(III)/Er(III), convection leads to preferential deposition of copper. Stirring also has a tendency to 'smear' the deposited species across the electrode surface, thus resulting in localised deposition or removal of the precursor metals.

Y-Ba-Cu-O films have also been prepared by Slezak et al. by deposition of the relevant metals from an aqueous hydroxide solution [8]. The results showed that the fabricated films had low superconducting volume fractions. However, a T_C of ~ 85 K was achieved.

The most prolific publisher in the field of electrochemical deposition of superconductors has been Bhattacharya et al. who have been publishing results continually since 1990 on many aspects of electrodeposition of mainly the thallium-based superconducting films. Films of $TlBa_2Ca_2Cu_3O_{9+\delta}$, $(Bi,Pb)_2Sr_2CaCu_2O_{8+\delta}$, and $YBa_2Cu_3O_{7-\delta}$ have been synthesised via several different routes. Firstly, the group produced superconductor precursor films of Bi-Pb-Sr-Ca-Cu on MgO and Y-Ba-Cu on ZrO by employing a constant potential approach [9,10]. After appropriate heat treatment, the superconducting films obtained displayed critical current densities of 500 A cm^{-2} and 4×10^3 A cm^{-2} at 4 K in zero field for (Bi,Pb)-2212 and Y-123 respectively. Tl-Ba-Ca-Cu precursor films, on MgO, were also produced by the same technique [11,12] which resulted in superconducting films of Tl-1223 with $T_{C,0}^*$ of 102 K and $J_{C,t}$ of 2×10^4 A cm^{-2} at 76 K. Also superconducting Tl-2223 films have been synthesised this way with $J_{C,t}$ of 3.2×10^4 A cm^{-2} at 76 K on a Ag substrate [13]. The difference here being that the films were electrodeposited using a pulsed potential technique (-4.0 V for 10 s and -1.0 V for 10 s using a Ag/AgNO$_3$ reference), and the as-deposited films were reacted in a two-zone furnace (figure 5.1) with an additional Tl source (a Tl-2223 pellet). Films synthesised without the additional Tl source showed Tl-1223 as the majority phase with a $J_{C,t}$ of 1×10^4 A cm^{-2} on Ag.

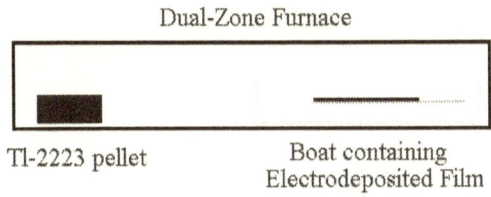

Figure 5.1 Schematic of dual-zone furnace. The Tl source and the films are maintained at different temperatures.

Bhattacharya et al. have also manufactured Tl-1223 superconducting films, on Ag substrates, doped with Ag [14]. Again a pulsed potential was applied and post deposition heat treatment was performed in a two-zone furnace. Transport critical current densities of 1.463×10^4 A cm^{-2} at 76 K and in zero field were achieved. By far the best transport properties were produced when Ag doped Ba-Ca-Cu films were produced via the a pulsed-potential electrochemical route and annealed in a two-zone furnace within an O$_2$ atmosphere [15]. The resulting films had $J_{C,t}$ of 4.42×10^4 A cm^{-2} at 76 K in zero field and an impressive 8.2×10^3 A cm^{-2} at 76 K in an applied field of 1 T. See table 5.1 for a summary of electrochemical studies performed as of July 1996.

The reviewed literature yields very little information concerning the experimental considerations and their effects upon the deposition procedure. Applied potential regimes include both

---

$^*T_{C,0}$ is the temperature at which the resistance goes to zero.

constant and pulsed-potential techniques but differences that arise from these methods are not discussed. All experiments were performed with no form of solution agitation (except in [7]), and solutions comprised of hydrated nitrate salts in either DMF, or the more commonly used, DMSO. The effects of temperature stability [15] and solution $Cu(NO_3)_2H_2O$ content upon the electrodeposition process [7] were mentioned briefly.

Chapter 5: Experimental Considerations for the Electrodeposition of....

Research Group	Compound	Electrochemical Technique, Materials, Solvent	Cell-type, Reference, Counter, Substrate	Deposition Rate ($\mu m\ min^{-1}$)	Transition Temperature, T_C (K)	Critical Current Density, $J_{C,t}$ (A cm^{-2})	Ref.
Materials Research Laboratory and School of Chemical Sciences, University of Illinois, Illinois.	Ba-Ca-Cu	Cyclic voltammetry Barium Hydroxide Solution	3-electrode Ag/AgCl Pt Cu	-	-	-	[2]
Allied-Signal, Inc., Corporate Technology, Morristown, New Jersey	$Bi_2Sr_2CaCu_2O_{8+\delta}$ $(Bi,Pb)_2Sr_2CaCu_2O_{8+\delta}$	Constant potential Dehydrated nitrate salts DMSO	3-electrode Ag/Ag$^+$ Pt MgO	-	85	300-400 at 40 K, 0 T	[3]
Solar Energy Research Institute, Golden, Colorado	$(Bi,Pb)_2Sr_2CaCu_2O_{8+\delta}$ $YBa_2Cu_3O_{7+\delta}$ $TlBa_2Ca_2Cu_3O_{9+\delta}$	Constant potential Hydrated nitrate salts DMSO	3-electrode Ag/AgNO$_3$ Pt Ni, MgO, ZrO, Al$_2$O, SrTiO$_3$	~1.0	Y123/Ni: 74 K Y123/MgO: 78 K Y123/ZrO: 91 K Y123/SrTiO$_3$: 80 K (Bi,Pb)2212/Al$_2$O$_3$: 62 K Tl1223/SrTiO$_3$: 102 K	Y123/ZrO: 4 × 10^3 at 4 K, 0 T Y123/ZrO: 360 at 77 K, 0 T Y123/SrTiO$_3$: 5.2 × 10^3 at 4 K, 0 T Y123/MgO: 3.96 × 10^3 at 4 K, 0 T (Bi,Pb)2212/MgO: 500 at 4 K, 0 T Tl1223/SrTiO$_3$: 2 × 10^4 at 76 K, 0 T Tl1223/SrTiO$_3$: 5 × 10^3 at 76 K, 1 T	[9-12] [14,15]
	$Tl_2Ba_2Ca_2Cu_3O_{10+\delta}$	Pulsed potential Hydrated nitrate salts DMSO	3-electrode Ag/AgNO$_3$ Pt MgO, Ni, Ag	~1.0	Tl1223/Ag: 110 K Tl2223/Ag: 120 K	Tl1223/Ag: 1.463 × 10^4 at 76 K, 0 T Tl2223/Ag: 3.2 × 10^4 at 76 K, 0 T Thalliniated Ba-Ca-Cu-O: 4.42 × 10^4 at 77 K, 0 T	[13]
Energy Studies Laboratory, Department of Physics, Shivaji University, India	$Bi_2Sr_2Ca_2Cu_3O_{10+\delta}$ $DyBa_2Cu_3O_{7+\delta}$	Constant potential Hydrated nitrate salts Water	3-electrode SCE Graphite MgO	-	91 K	-	[4,5]
Department of Mechanical Engineering and Energy Processes and Physics, University of Carbondale, Illinois	$YBa_2Cu_3O_{7+\delta}$ $ErBa_2Cu_3O_{7+\delta}$	Pulsed potential Hydrated nitrate salts DMF	3-electrode Ag/AgCl Pt Ag coated ZrO$_{2+\delta}$, SrTiO$_3$, CaTiO$_3$	~0.06	Y123/ZrO$_2$: 93 K Er123/CaTiO$_3$: 80 K	-	[6,7]

Table 5.1 A summary of electrochemical studies performed on superconducting films.

5.3 Experimental Considerations

There are many factors that affect the deposition process and must therefore be considered before attempting to fabricate superconductor precursor films. A major problem in determining these considerations is the lack of information in the literature. DMSO is not a particularly common solvent and research into the behaviour of the individual metals in DMSO is limited. Details of the deposition of several metals simultaneously is even scarcer. The following section explains the factors to be considered and discusses the effect they have upon the deposition process, especially reproducibility. These factors include:

1. applied potential range;
2. substrate preparation and substrate type;
3. water and oxygen content of the electrochemical solution;
4. deposition technique applied, i.e. normal, stirred, and ultrasonic;
5. temperature stability;
6. solution ion depletion;
7. total metallic ion concentration;
8. relative metallic ion concentration.

Following this the reproducible electrodeposition of Bi-Sr-Ca-Cu superconductor precursor films is described.

5.3.1 Selecting the Correct Applied Potential

In selecting the correct potential for co-deposition the stability and the behaviour of the constituent components must be taken into account. The potential must be sufficient to reduce all the included metal ions, e.g. Tl^+, Pb^{2+}, Sr^{2+}, Ca^{2+}, and Cu^{2+}, but not so large as to significantly breakdown the solvent used. Moreover, an appropriate rate of deposition must be sustained. Figure 5.2 is a cyclic voltammagram of a Tl-Pb-Sr-Ca-Cu solution used in the deposition of thallium-based precursor films. The solution comprised of 12.0 mM $TlNO_3$, 2.0 mM $Pb(NO_3)_2$, 225.0 mM $Sr(NO_3)_2$, 450.0 mM $Ca(NO_3)_2H_2O$, and 16.0 mM $Cu(NO_3)_2H_2O$ dissolved in DMSO. The applied potential was swept from 0 to -3.0 V (versus Ag psuedo reference) at a rate of 100 mV s^{-1}, and the cell was regulated at a temperature of 30 °C. The assignment of the different features observed in the cyclic voltammogram was decided from the cyclic voltammograms obtained for the different elements presented in the previous chapter. From figure 5.2 it can be observed that potentials < -2.20 V (versus a Ag wire psuedo reference and at 30 °C) must be applied if all the ions are to be reduced simultaneously. In practice, an applied potential of -3.25 V to -3.50 V was used, which led to a satisfactory growth rate. It can be seen also that in this range breakdown of the DMSO (< -2.50 V) also occurs but EDS analysis showed that significant amounts of sulphur were *not* incorporated into the film.

If the height of the current peaks are compared to the ionic concentrations in the solution, it is immediately clear that the peak due to the reduction of Sr and Ca cations is far too small to be mass transport limited. This is probably due to the passive nature of Sr and Ca deposits, preventing further deposition from occurring.

The choice of applied potential is also important when considering efficient use of the materials because of the disparity in the constituents' redox potentials. This arises because the difference between the applied potential and the redox potential of a particular metallic cation is directly related to the growth rate (and the morphology) of the film, i.e. the greater the difference the faster the deposition occurs, assuming that the process does not become mass transport limited (see reference [18] in chapter 4). This fact means that, because of the large difference in redox potentials between Tl^+, Pb^{2+}, Cu^+ and Sr^{2+}, Ca^{2+}, relatively large amounts of Sr and Ca nitrates must be dissolved in the solution compared to Tl, Pb, and Cu nitrates in order to produce films with significant amounts of Sr and Ca present. Obviously, the applied potential could be increased but this two has its associated problems.

Chapter 5: Experimental Considerations for the Electrodeposition of....

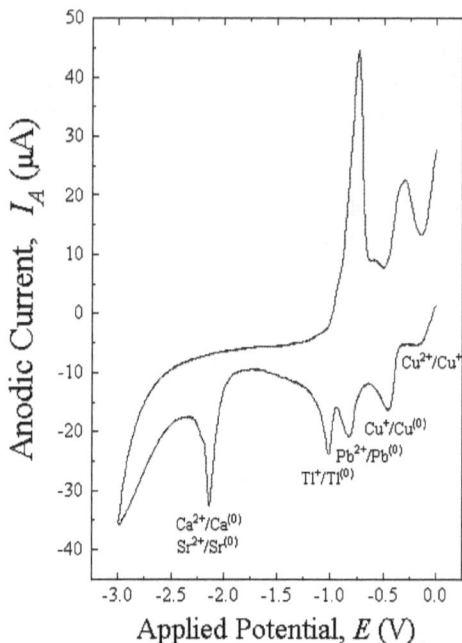

Figure 5.2 *A cyclic voltammagram of a solution comprising of 12.0 mM TlNO$_3$, 2.0 mM Pb(NO$_3$)$_3$, 225.0 mM Sr(NO$_3$)$_2$, 450.0 mM Ca(NO$_3$)$_2$, and 16.0 mM Cu(NO$_3$)$_2$ dissolved in DMSO. The experiment was performed in a dry box at 30°C with a sweep rate of 100 mVs^{-1}. The electrode area was ~ 0.4 cm^2.*

The magnitude of the applied potential affects the morphology of the films produced. Figure 5.3 shows the effect of applied potential upon the stoichiometry of electrodeposited Ba-Ca-Cu films. Note that as the potential becomes more negative the film content of barium and calcium is reduced compared to copper, *or* the copper content of the film's produced increases with respect to barium and calcium. This is due to the difference in potentials between the redox potentials and the applied potential of the relevant cations resulting in differences on the individual deposition rates.

If the applied potential used to deposit precursor films is too large, however, then poorly adhered films would be produced with very high void fractions. This can be seen in figure 5.4 which is an SEM of a Ba-Ca-Cu film deposited at an applied potential of -5.0 V versus a Ag psuedo reference at 18 °C (where DMSO breakdown occurs at applied potentials < -3.2 V). There is evidence that a high void fraction would result in the superconducting film having inferior current carrying properties as opposed to those produced from heat treated films with dense solid morphologies [17]. The transport critical current density is directly related to the superconducting volume fraction which in turn is related to the film void space. An attractive feature of electrodeposition is the ability to grow thick films in order to achieve good transport critical currents as opposed to high transport critical current densities.

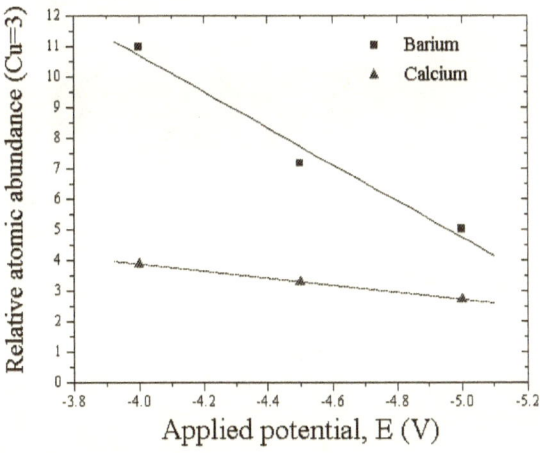

Figure 5.3 A plot of film content versus applied potential (versus a Ag psuedo reference). Deposition was performed at 18 °C in a dry box. The solution composed of 10.0 mM $Ba(NO_3)_2$, 10.0 mM $Ca(NO_3)_2$, and 5.0 mM $Cu(NO_3)_2$, dissolved in DMSO. The solution was renewed between depositions.

Figure 5.4 An SEM of a Ba-Ca-Cu film deposited at -5.0 V (versus an Ag psuedo reference) from a solution composed of the same nitrate salt concentrations as the solution described in figure 5.3. (magnification: × 10)

5.3.2 Temperature Stability

To achieve reproducibility over a long period of time the temperature must be maintained constant. A variation of ~ 10 °C is sufficiently large to severely affect the reproducibility of the process over the year. This effect can be clearly seen in figure 5.5 for the deposition of Bi-Sr-Ca-Cu films deposited at temperatures between 30 °C and 50 °C. Figure 5.5a shows that the total charge transferred during each experiment, and therefore the film thickness, increased as temperature increased. The increase of charge with increasing temperature can be attributed to an increase in the mass transfer rate. Figure 5.5b shows the effect of temperature upon the stoichiometry of the deposited films. The observed variation of film stoichiometry is probably due to the changing diffusion constants of the metallic cations in the deposition solution with temperature and cation radius: $D \propto \sqrt{T}/r$. This relation is derived from an equation known as the Stokes-Einstein relation which related the diffusion constant of a particle to it's radius (r), solvent temperature (T), and the solvent viscosity (η).

$$D = \frac{k_B T}{6\pi \eta r} \qquad \{5.1\}$$

Chapter 5: Experimental Considerations for the Electrodeposition of....

The viscosity, η, can be expressed as,

$$\eta = \frac{1}{3\sigma}\sqrt{\frac{4mk_bT}{\pi}} \rightarrow \eta \propto \sqrt{T} \qquad \{5.2\}$$

where m is the mass of one of the solvent molecules and σ is the collision cross-section for the solvent. Combining 5.1 and 5.2 yields the dependence,

$$D = \frac{\sigma}{4}\left(\frac{k_B}{\pi m}\right)^{1/2}\frac{\sqrt{T}}{r} \rightarrow D \propto \frac{\sqrt{T}}{r} \qquad \{5.3\}$$

Hence, the changing mass transfer rate implied by figure 5.5a is confirmed by theory.

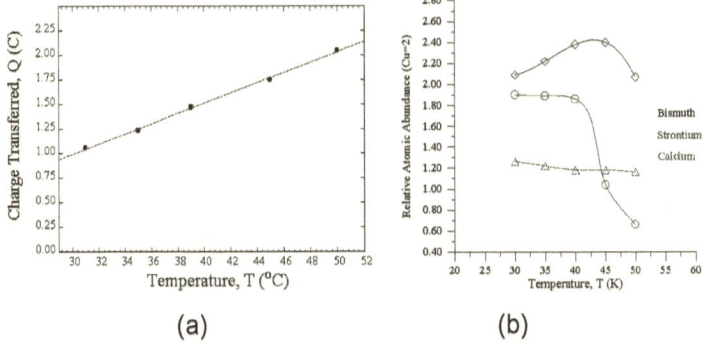

Figure 5.5 *The variation of: (a) total charge transferred versus temperature, and (b) film stoichiometry versus temperature for electrodeposited Bi-Sr-Ca-Cu films. The deposition solution comprised 5.0 mM Bi(NO₃)₃, 5.0 mM Sr(NO₃)₂, 6.0 mM Ca(NO₃)₂H₂O, and 7.0 mM Cu(NO₃)₂H₂O dissolved in DMSO. The solution was maintained at 30 °C and a constant potential was applied at -3.25 V for 1800 s. The solution was renewed for each film.*

On a smaller scale, the deposition process can be affected, during shorter time scales (~ 10 s), by thermal convection currents within the solution due to temperature variations throughout the cell. Both the above situations can be overcome by mounting the electrochemical cell in a dry block heater.

5.3.3 The Effect of Water

The problem with water is that the amount found in the materials used is variable, depending upon the way they are stored, and this can lead to difficulties in preparing a solution with a specific metal concentration. Water also alters the chemistry of the complexes formed and of the deposits produced. Because of the high applied potentials used water is also reduced during the deposition procedure. When water is reduced hydrogen gas is produced at the face of the substrate. This may interfere with the process by causing small convective disturbances.

As mentioned previously (§ 4.2.7.6) oxygen can react with protons in the solution producing hydrogen peroxide followed by water, again affecting the deposition process. Oxygen can be removed from the deposition bath by degassing it in an ultrasonic bath.

5.3.4 Metallic Ion Depletion

Owing to the disparity of the reduction potentials between the different metallic ions involved the amount of cations of Sr, Ca, Ba compared to Tl, Pb, Hg, Cu, and Bi dissolved in the deposition bath is large. Therefore, there is a problem with depletion of the more noble metals, i.e. as time passes the solution content of, say Tl^+, diminishes and, hence, the film content is not reproducible. Figure 5.6 shows this by plotting the variation in film composition of Tl-Ba-Ca-Cu films deposited from the same solution. The films were deposited from a solution comprising 5.0 mM $TlNO_3$, 200.0 mM $Ba(NO_3)_2$, 350.0 mM $Ca(NO_3)_2H_2O$, and 12.0 mM $Cu(NO_3)_2H_2O$ dissolved in DMSO. The films were deposited at 18 °C from the same solution. The applied potential was -4.0 V, 10 s and -1.0 V, 10 s cycled for 600 s. The depletion can be accounted for as follows. During the deposition of sample 1 (in figure 5.6) a total charge of ~ 1 C was transferred of which ~ 0.3 C can be attributed to the deposition of thallium (calculated from the film stoichiometry). This means that ~ 2×10^{18} thallium cations were removed from the 10 ml solution, which is equivalent to the solution thallium concentration changing from 5.0 mM to 4.6 mM - a reduction of ~ 10 % which is sufficiently large to account for the changing film stoichiometry.

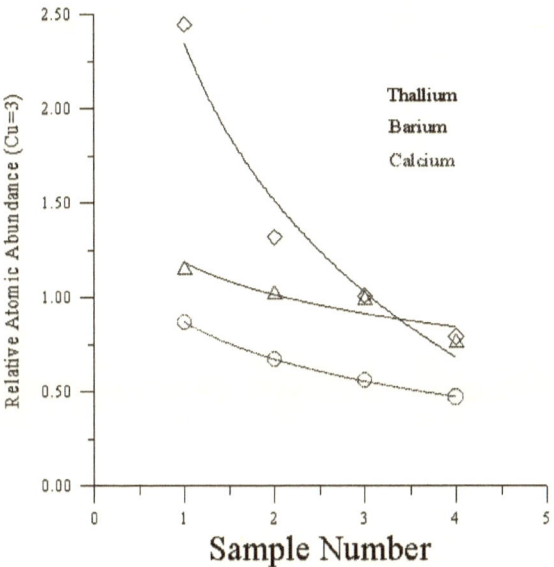

Figure 5.6 Sample number versus film composition for films deposited from the same solution. The solution was composed of 5.0 mM TlNO$_3$, 200.0 mM Ba(NO$_3$)$_2$, 350.0 mM Ca(NO$_3$)$_2$, and 12.0 mM Cu(NO$_3$)$_2$ in DMSO. Deposition was performed in a dry box at 18 °C, with a pulsed applied potential of -4.0 V, 10 s and -1.0 V, 10s. The volume of solution used for each film was 10 ml. The solution was not renewed between experiments.

To ensure reproducible solution characteristics the deposition solution was changed between experiments in order to achieve reproducible film stoichiometries. Also, the effect could easily be eradicated by simply working with much larger solution volumes - the experiments herein were limited by the working space and materials. This result also indicates that the stoichiometry of the deposited films was dependent upon the deposition period, suggesting that the distribution of the metals throughout the films is not constant.

5.3.5 Substrate Selection

The substrate used depends upon the application in mind. However, for all applications it is desirable to have well aligned superconducting films. To achieve this the lattice mismatch, i.e. the difference between the substrate and the superconductor lattice parameters, must be minimised. Ag has a reasonable lattice mismatch and has been used extensively because of the relatively low costs compared to more suited substrates such as

$SrTiO_3$, MgO, ZrO_2, and $CaTiO_3$. The substrate selection has a considerable effect upon the superconducting properties of the deposited films, in particular, the transport properties [16]. Herein, Ag is used not only because of its low costs but also because of its flexibility - important if a continuous process is to be developed.

5.3.6 Electrodeposition: Silent, Stirring, and Ultrasonic

The deposition rate and the film morphology can be significantly controlled by employing a variety of electrochemical techniques. The main difference between the different approaches is the way in which they affect the extent of the diffusion layer through mechanical agitation. In this work, the deposition performed is *silent*, i.e. no deliberate agitation. However, the influence of stirring the solution during deposition, and performing deposition in an ultrasonic field, was investigated to observe their effect upon growth rate and morphology.

5.3.6.1 Silent Deposition

Apart from experiments performed in this section all the deposition in this work is *silent*, i.e. no solution agitation involved during the deposition process. All influential factors upon the process, as described previously, are removed or controlled. This means that in all experiments,

- the solution was prepared from dehydrated materials, including the solvent.
- a fixed geometry cell was used.
- a planar Pt gauze forms the counter electrode, and a section of Ag wire was used as a psuedo reference. Both are suspended vertically.
- deposition was performed in a dry argon atmosphere.
- the cell was mounted in a dri-block® heater DB-2A (Techne).
- the solution was renewed and degassed between experiments.

This approach yields porous films, and a very high standard of reproducibility. The growth rate, though, is relatively low.

5.3.6.2 Stirred Deposition

Stirring is integrated into the process by adding a magnetic stirrer to the experimental arrangement. The philosophy behind this addition is that stirring increases the mass transport, and hence, the deposition rate, as well as making the current distribution more uniform. Forced convection leads to a thin layer of solution next to the electrode, within which it is assumed that only diffusion occurs - the diffusion layer of thickness δ. If the value of δ is constant over the whole electrode surface then the electrode is uniformly accessible to electroactive species that arrive from bulk solution.

There is a negative side though. The stirring produces *cylindrical* convection currents within the solution, but the working substrate is *planar*. This causes eddy currents to form at and near the substrate surface which significantly affects the concentration homogeneity close to the surface, i.e. δ is not constant over the electrode surface due to the different geometries involved. The effect upon deposition is that the films have a streak running down their length where the eddy currents have limited deposition (figure 5.6).

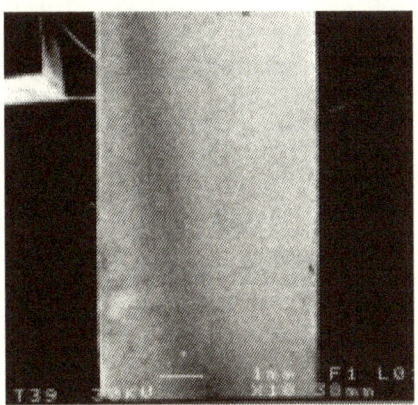

Figure 5.6 SEM of Tl-Pb-Sr-Ca-Cu film deposited whilst stirring the deposition solution. The film was deposited from a solution described earlier. A magnetic stirrer was incorporated into the procedure to induce convection to the solution. The angular velocity of the stirring bar was ~ 5 rad s^{-1}. Deposition was caused by applying a potential of -4.0 V, 10 s and -1.0 V, 10 s cycled for 600 s. The temperature was maintained at 18 °C. Note the 'streak' running down the length of the film due to the eddy currents present close to the electrode.(magnification: × 10)

5.3.6.3 Deposition in the Presence of an Ultrasonic Field

One of the key parameters that affects the critical current density, $J_{C,t}$, of the films is the relative density, i.e. the actual coverage of the superconducting film on the substrate [17]. In films manufactured via an electrochemical stage there can be as high as 90 % void space (see § 5.4.2) which does not lead to high superconductor coverage once the precursor has been heat treated to yield the correct superconducting phase. Ultrasonic irradiation has previously been applied in metal plating and has been shown to keep the electrodes clean, the solution degassed, as well as improving the mass transport to the electrode surface by disturbing the diffusion layer preventing ion depletion [18]. Other benefits, which have been observed in electrochemical polymerisation [19], include improved adhesion to the working electrode of the deposit, enhanced hardness of the deposit, while the process itself has a lower deposition current and a higher deposition rate. These effects can be attributed to the actions of micro-agitation and cavitation produced by the ultrasonic field [20].

The effect of an ultrasonic field, applied during deposition, was investigated to observe the changes in the films produced, in particular, the films' density, or void space. The ultrasound was incorporated into the process by suspending the cell in an ultrasonic bath (Ultrawave: 100 W, 30 kHz) (see figure 5.7). All other necessary steps were taken to ensure a reproducible procedure.

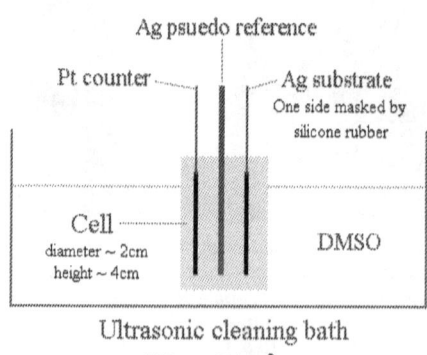

Figure 5.7 *Experimental set-up for electrodeposition in the presence of an ultrasonic field.*

Deposition of Tl-Pb-Sr-Ca-Cu films was attempted onto one side of Ag foil from a solution comprising 3.2 mM TlNO$_3$, 1.8 mM Pb(NO$_3$)$_2$, 25.0 mM Sr(NO$_3$)$_2$, 54.0 mM Ca(NO$_3$)$_2$H$_2$O, and 6.0 mM Cu(NO$_3$)$_2$H$_2$O dissolved in DMSO. Film growth was carried out at 18 °C, with and without an ultrasonic field, with the application of a pulsed potential of: -4.0 V, 10 s and -1.0 V, 10 s, this was repeated 25 times.

Figure 5.8 shows the current/time transients for films deposited with and without the presence of an ultrasonic field. It can be clearly seen that the effect of the ultrasonic irradiation upon the mass transfer rate is to increases it four-fold. This effect is derived from the asymmetric collapsing of the cavitation events in the vacinity of the working electrode, which in turn causes jets of solution to impinge onto the electrode. Note also that the average current increases with time for an ultrasonically deposited film - this is due to an increase in temperature of 7 °C over the deposition period, again caused by ultrasonic processes. The influence of the ultrasound means that during the 500 s deposition period a charge of 1400 mC is passed as opposed to 246 mC for a film produced in the absence of ultrasound. This is a favourable effect as in practice precursor films could be synthesised rapidly as compared to approaches attempted thus far. The inset of figure 5.8 shows a typical pulse transient produced during the deposition process. The shape of the transients can be explained as follows. The co-deposition of the five components is difficult because of the passive nature of Ca metal in the films. To overcome this problem the applied potential is pulsed so that, during the -1.0 V part of the pulse, a conductive layer of Cu, Tl, and Pb (CTP) is deposited. The CTP layer assists in the co-deposition, performed at -4.0 V, of all the components. This cycle of co-deposition/CTP deposition maintains the deposition rate, whereas, if a constant potential of -4.0 V was applied to cause co-deposition then the passive nature of the deposit hinders further deposition.

EDS analysis of the films indicated that the composition also depended significantly on whether an ultrasonic field was present or not. Films deposited in the absence of an ultrasonic field had an average composition of Tl:Pb:Sr:Ca:Cu = 1.0:0.5:1.1:1.24:3.0, whereas films produced in an ultrasonic field had an average composition of Tl:Pb:Sr:Ca:Cu = 0.8:0.8:0.2:0.2:3.0. The main difference is the diminished abundance of, in particular, Sr and Ca. This change is probably due to the fact that ultrasonic irradiation keeps the electrodes clean during the deposition process. This suggests that poorly adhered material is removed throughout the procedure, also suggesting that the current transfer is not completely efficient.

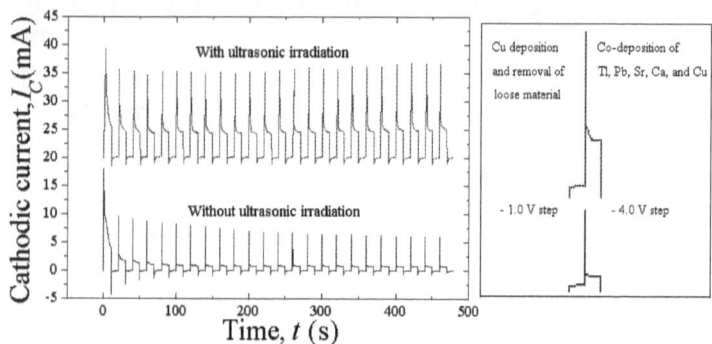

Figure 5.8 Current versus time plot for normal and ultrasonically electrodeposited Tl-Pb-Sr-Ca-Cu films. The films were deposited from a solution comprising 3.2 mM TlNO$_3$, 1.8 mM Pb(NO$_3$)$_2$, 25.0 mM Sr(NO$_3$)$_2$, 54.0 mM Ca(NO$_3$)$_2$H$_2$O, and 6.0 mM Cu(NO$_3$)$_2$H$_2$O dissolved in DMSO. The initial temperature was 18 °C and deposition was performed with and without ultrasonic agitation of the solution. The applied potential was -4.0 V, 10 s and -1.0 V, 10 s cycled for 500 s. The area of the electrode was ~ 0.4 cm^2. The plot for the film deposited in the presence of ultrasound has been shifted up by 20 mA for clarity.

SEM micrographs of the films show a more important difference between the deposition techniques. Micrographs reproduced in figures 5.9a, b, and c show how the surface morphologies of the films evolve from clumps of deposit to, what appears, dense homogeneous films. Figure 5.9a displays the clumpy morphology of films synthesised in the absence of ultrasound, the films not having a high level of morphological homogeneity, whereas figure 5.9b, in which the film was produced in an ultrasonic field with a initial solution temperature of 18 °C, the films' morphology has improved drastically to produce a more homogeneous appearance. The improvement in film densification and homogeneity was furthered when synthesis occured in a solution with an initial temperature of 25 °C (figure 5.9c). Here microcracking was also observed.

By assuming that the process is 100 % efficient, and that, ideally, a dense phase was formed it was possible to predict the volume of a non-porous film. By comparing this value to the actual volume of the deposits it was possible to obtain a measure of the film void percentage. Figures 5.10a and b are cross-sectional SEM micrographs of films manufactured from solutions with different initial temperatures in the presence of an ultrasonic field. At 18 °C, 1400 mC was transferred resulting in a film approximately 2.7 µm thick with an area of ~ 0.4 cm^2. Given that the predicted value is 1.5 µm thick a value of [(2.7-1.5)/2.7]×100 = 45 % is obtained for the film void percentage, almost half of that obtained previously for normally deposited films. This value, however, increases for films deposited at higher initial solution temperatures, possibly indicating that an ideal situation exists for producing dense films via an ultrasonic electrochemical deposition process. It should be pointed out that it is believed that the process is not 100 % efficient, because of loss of powdered material, and so further investigation is required to determine the process efficiency and its dependence on solution temperature.

In this section it has been shown that by incorporating an ultrasonic field, by means of performing all deposition in a cell placed in an ultrasonic cleaning bath, into the electrochemical deposition set up superconductor precursor films, with superior homogeneity and density, can be synthesised. Attempts to increase the amount of Sr and Ca in the as-deposited films, however, were not successful possibly due to the poor adherence of these elements to the deposited film. As the solution content of the Sr and Ca nitrate salts was increased the morphology of the resulting films was very uneven. Additionally, it was found that if the total cation concentration exceeded ~ 100 mM then deposition was severely hindered, possibly due to a too greater deposition rate. Before further progress can be made in the ultrasonic electrodeposition of superconductor precursor films, a greater understanding of the mechanical effects of ultrasonic irradiation upon films evolution is required.

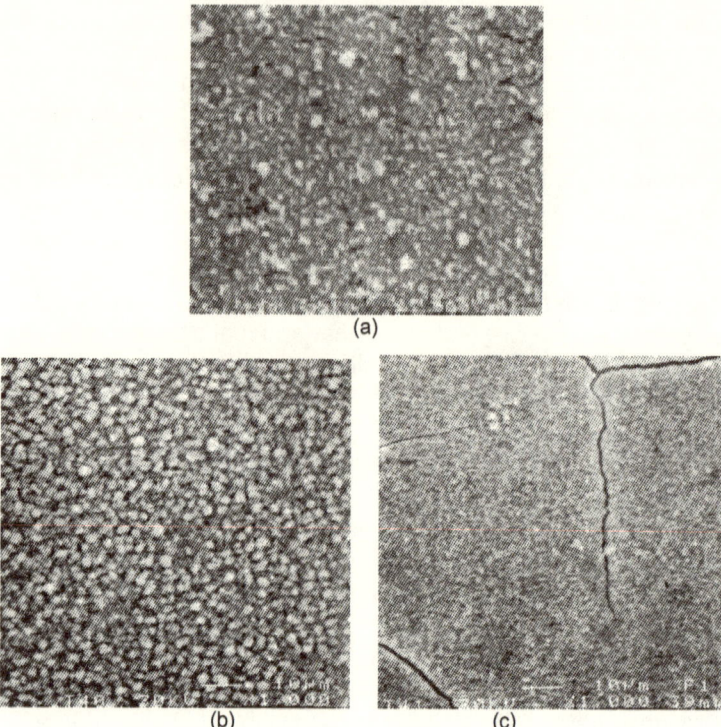

Figure 5.9 SEM's of films electrodeposited from a solution comprising 3.2 mM $TlNO_3$, 1.8 mM $Pb(NO_3)_2$, 25.0 mM $Sr(NO_3)_2$, 54.0 mM $Ca(NO_3)_2H_2O$, and 6.0 mM $Cu(NO_3)_2H_2O$ dissolved DMSO. An applied potential of -4.0 V, 10 s and -1.0 V, 10 s is cycled 25 times. The SEM's are films deposited using (a) normal deposition at 18 °C, (b) ultrasonic deposition at 18 °C, and (c) ultrasonic deposition at 25 °C. (magnification: ×1000)

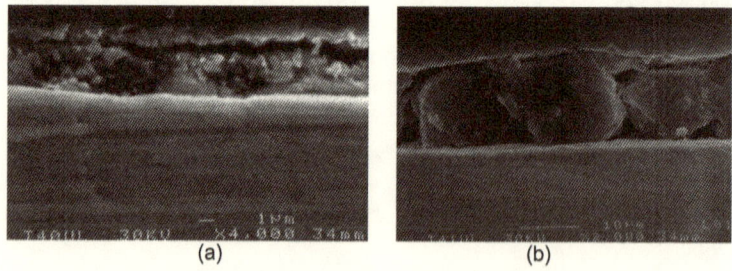

Figure 5.10 Cross-sectional SEM of ultrasonically electrodeposited films, deposited at (a) 18 °C (magnification: ×4000 and (b) 25 °C (magnification: ×2000).

5.4 Electrodeposition of Bi-Sr-Ca-Cu Films

Hitherto, the experimental considerations have been identified and discussed. In this section the above ideas are brought together in the reproducible synthesis of quality Bi-Sr-Ca-Cu films.

5.4.1 Solution Optimisation

Three groups of solutions were prepared in the dry box, each group varying the relative concentration of a particular metal whilst leaving the amount of the others constant. For example, the first group of solutions had a range of concentrations for copper nitrate, whilst bismuth, strontium, and calcium remained constant. The second group varied strontium, and the third calcium. Bismuth was kept constant throughout all the solutions as a point of reference. A film was fabricated from each of the solutions, at 30 °C, and after drying was analysed using EDS to determine their composition.

Figures 5.11a and 5.11b show the film composition, and the total charge transferred during deposition, respectively, as a function of solution copper concentration. Because copper is a good conductor (resistivity = 1.67 $\mu\Omega$ cm at 20 °C) it is easily deposited and this is reflected in the increase of copper (relative to bismuth) in the as-deposited film. The total charge transferred also increases indicating that the films were thicker as copper content increased, again reflecting the ease by which copper films can be grown. The relative film content of strontium and calcium generally decreases as the solution copper content increases.

The series of films deposited from solutions with increasing strontium content, however, display a very different dependence because of the relatively poor conduction properties of strontium (resistivity = 23.0 $\mu\Omega$ cm at 20 °C). From the total charge transferred versus solution strontium concentration (figure 5.12a) it can be observed that the film thickness increases very slightly with increasing strontium concentration. In fact, at higher solution strontium concentrations the deposition process was slowed by the low conductivity of the films (therefore thinner films) due to the large amount of poorly conducting strontium contained within the films. Further evidence for the limiting effects of strontium content upon film growth was observed in the film stoichiometries (figure 5.12b). The relative film content of calcium and copper remain generally the same within the solution strontium concentration limits applied to the set of experiments. The film strontium content increases slightly but then was observed to decrease at higher solution concentrations, indicating the difficulty in fabricating films containing large amounts of strontium.

Finally, the series of films deposited from solutions with increasing calcium content clearly show the effects upon the deposition process from a constituent that forms passive layers. Figure 5.13a shows how the total charge transferred, and hence film thickness, drops rapidly with increasing solution calcium concentration - the total charge transferred falling to a third of its initial value when the solution calcium concentration is doubled. Figure 5.13b shows the variation of film stoichiometry with solution calcium concentration. The film calcium content begins to increase but quickly saturates, indicating a limit to the amount of calcium that can be included in films of this type. Unexpectedly, the film strontium content falls, but interestingly the copper content increases. The increase in copper content can probably, again, be attributed to the good electrical properties of copper metal.

From this data it can be seen that not only can the desired 2212 ratio be achieved, but it seems to be an favourable ratio for this system. For example, because of the response from increased calcium and strontium, it would be difficult to grow films with substantially greater amounts of these two elements.

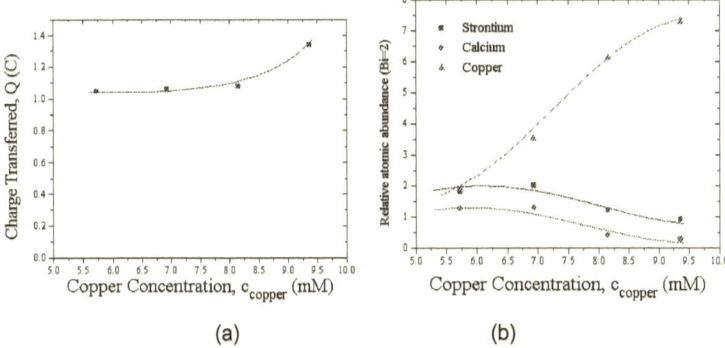

Figure 5.11 (a) The total charge transferred versus solution copper content. (b) Film composition versus copper concentration. Films were deposited in a dry box at 30 °C from a deposition bath composing of 6.0 mM $Sr(NO_3)_2$, 7.0 mM $Ca(NO_3)_2$, 5.0 mM $Bi(NO_3)_3$, and differing amounts of $Cu(NO_3)_2$ all dissolved in DMSO. The applied potential was -3.25 V versus a Ag pseudo reference and held for 1800 s. The area of the films was ~ 0.35 cm^2.

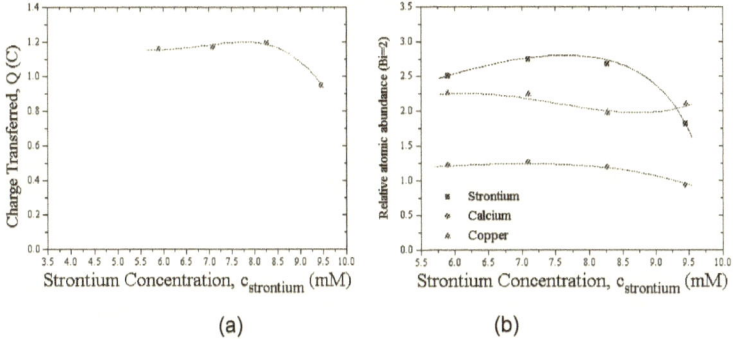

Figure 5.12 (a) The total charge transferred versus solution strontium content. (b) Film composition versus copper concentration. Films were deposited in a dry box at 30 °C from a deposition bath composing of 6.0 mM $Cu(NO_3)_2$, 7.0 mM $Ca(NO_3)_2$, 5.0 mM $Bi(NO_3)_3$, and differing amounts of $Sr(NO_3)_2$ all dissolved in DMSO. The applied potential was -3.25 V versus a Ag pseudo reference and held for 1800 s. The area of the films was ~ 0.35 cm^2.

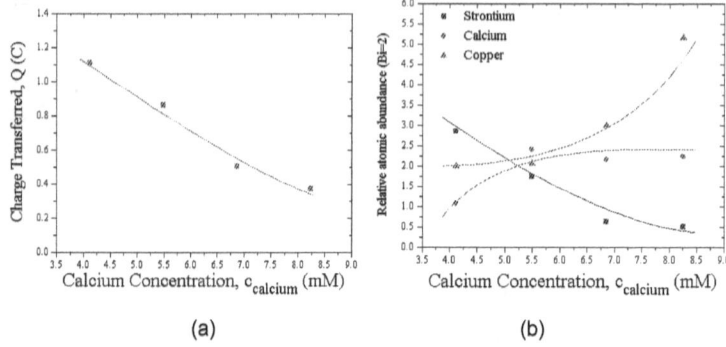

Figure 5.13 (a) The total charge transferred versus solution calcium content. (b) Film composition versus copper concentration. Films were deposited in a dry box at 30 °C from a deposition bath composing of 6.0 mM $Sr(NO_3)_2$, 7.0 mM $Cu(NO_3)_2$, 5.0 mM $Bi(NO_3)_3$, and differing amounts of $Ca(NO_3)_2$ all dissolved in DMSO. The applied potential was -3.25 V versus a Ag pseudo reference and held for 1800 s. The area of the films was ~ 0.35 cm^2.

As well as the film compositions varying with solution composition, the film morphology also varies significantly. Figures 14a, b, and c show how the film morphology evolves with increasing copper, strontium, and calcium content, respectively. All the deposits appeared dark brown in colour after drying at 200 °C. As the solution copper content increased the grouped islands of deposit became smaller giving the films a cracked appearance (figure 14a). The deposits were increasingly less well adhered to the substrate with increasing copper content and material easily fell away from the substrate. This is due to the difference between the reduction potential of the copper and the potential applied to achieve co-deposition of the constituent metals. This causing poorly adhered deposits.

As the solution strontium concentration was increased the film morphology evolved in a similar manner to the films deposited from solutions with increasing copper concentrations. The grouped islands of deposit became smaller but more significantly than the varied copper films. Cracking also appeared in films deposited from solutions with relatively high strontium concentrations. The morphological film evolution of films deposited from solutions with increasing calcium content was almost exactly the same as those deposited from solutions containing increasing strontium concentrations.

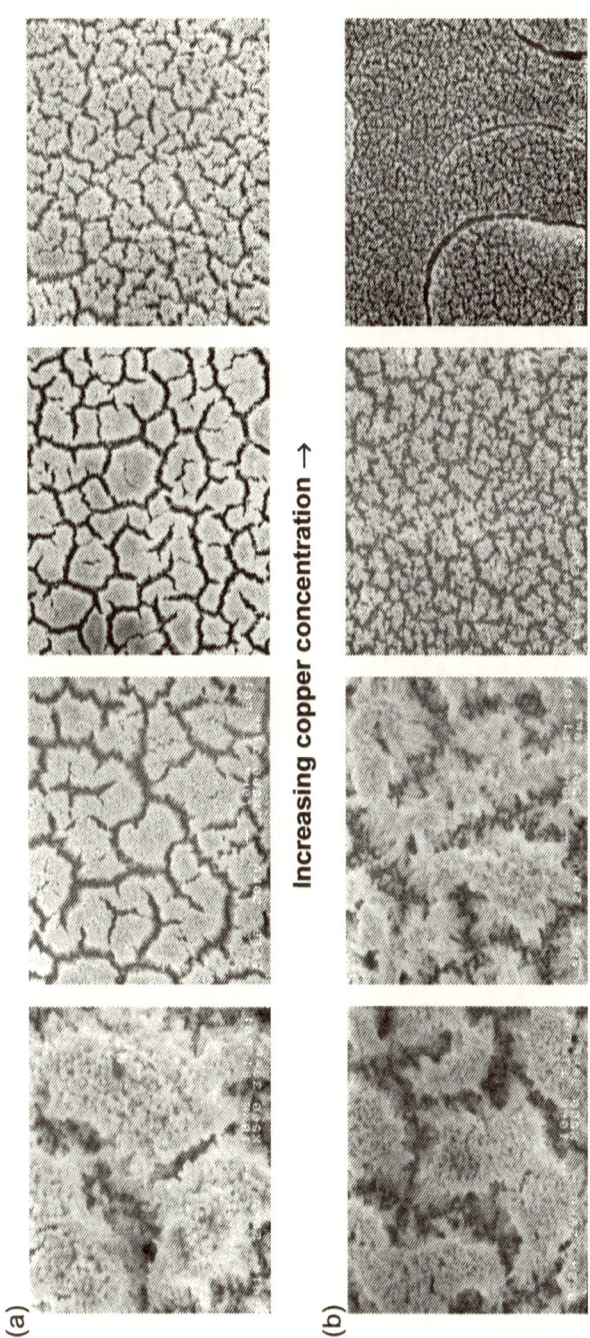

Figure 5.14 SEM's of films electrodeposited from solutions with increasing (a) copper, and (b) strontium concentration.

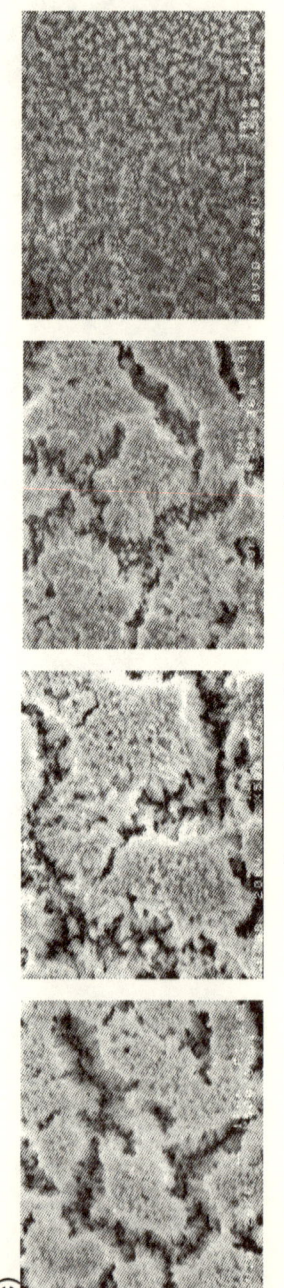

Increasing calcium concentration →

Figure 5.14c SEM's of films electrodeposited from solutions with increasing calcium concentrations.

5.4.2 Reproducibility and Heat Treatment

From the solution optimisation experiments it was found that a solution comprising of 6.0 mM $Bi(NO_3)_3$, 5.0 mM $Sr(NO_3)_2$, 5.0 mM $Ca(NO_3)_2H_2O$, and 7.0 mM $Cu(NO_3)_2H_2O$ dissolved in DMSO would yield films of the correct 2212 stoichiometry. The conditions were a temperature of 30 °C and an applied potential of -3.25 V held for 1800 s. A set of experiments were performed whereby pairs of films were deposited, on different days, from freshly prepared solutions to quantify the reproducibility of the procedure. Figure 5.15 shows the stoichiometry of the resulting films. The elemental composition of the precursor films does not vary by more than 10 % from film to film yielding a high standard of reproducibility.

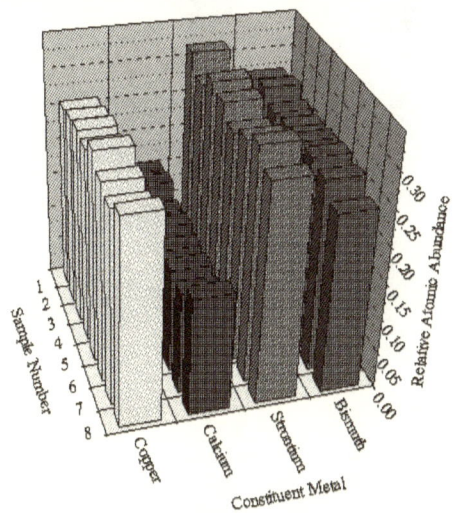

Figure 5.15 Reproducibility of electrodeposited Bi-Sr-Ca-Cu films deposited at 30 °C with an applied potential of -3.25 V (versus Ag pseudo reference) held for 1800 s from a deposition solution comprising 6.0 mM $Bi(NO_3)_3$, 5.0 mM $Sr(NO_3)_2$, 5.0 mM $Ca(NO_3)_2H_2O$, and 7.0 mM $Cu(NO_3)_2H_2O$ dissolved in DMSO.

Compositional determination of the films indicated that the ratio Bi:Sr:Ca:Cu within the films was typically 2.0:1.8:1.2:2.0 (see figure 5.16). This content did not vary by more than 10 % from point to point over the surface of any given film as measured by EDS. EDS also showed that the films contained oxygen but it was not possible to quantify the oxygen content.

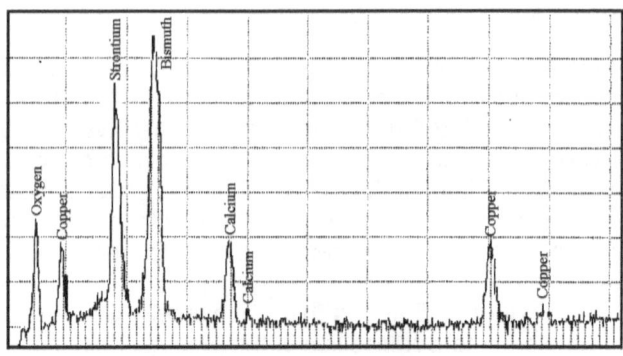

Figure 5.16 EDS of Bi-2212 electrodeposited film.

Figure 5.17 A cross-sectional SEM of an electrodeposited Bi-2212 precursor film. The film was deposited at 30 °C with an applied potential of -3.25 V (versus Ag pseudo reference) held for 1800 s from a deposition solution comprising 6.0 mM $Bi(NO_3)_3$, 5.0 mM $Sr(NO_3)_2$, 5.0 mM $Ca(NO_3)_2H_2O$, and 7.0 mM $Cu(NO_3)_2H_2O$ dissolved in DMSO. (magnification: ×1100)

Figure 5.17 is a SEM of the cross-section of a Bi-2212 precursor film. The figure indicates the porous nature of the deposited film, and yields a thickness value of ~ 48 μm. By comparing this to the predicted value for a dense single phase superconductor (~ 4 μm), obtained from the total charge transferred, an estimation of the void percentage can be calculated. In the case of Bi-2212 precursor films an extremely high void percentage of ~ 90 % is obtained.

The electrodeposited films were then subsequently heat treated in a tube furnace to obtain the superconducting $Bi_2Sr_2CaCu_2O_{8+\delta}$ phase (see Chapter 7).

5.5 Conclusions[†]

In this chapter the effect of experimental parameters upon the deposition process have been discussed in detail. From the results the main variables have been identified and their influence quantified. These effects include: the magnitude of the applied potential, deposition solution composition, deposition technique, and temperature stability. The applied potential not only affects the film morphology and quality, but more importantly the stoichiometry. All the parameters yield similar effects: temperature variations cause stoichiometric changes and therefore process reproducibility; deposition solution composition effects film quality and again reproducibility. The optimisation of the solution composition in order to yield films with a particular composition was found not to be trivial with the relationship of bath stoichiometry to film stoichiometry being a complex one. The reason for this complexity is probably due to the very different electrical conducting properties of the superconductor constituents. Elements with low electrical conductivity hinder the deposition process whereas elements with high electrical conductivity assist the deposition process.

Deposition technique undoubtedly has the most dramatic effect on the film structure and quality. Normal deposition causes porous films (90 % void space) with dendritic profiles, stirring produces streaky uneven films, and ultrasonic deposition results in dense, low void space (~ 45 %), films. However, these different approaches only influence the films' large and small scale morphologies via changes in the deposition rate. In fact, all the experimental parameters have simple effects on the fabrication procedure and are easily controlled with careful experimentation. Water, and oxygen are not so easily controlled from experience and have a more subtle effect on the process. The dendritic growth caused by the presence of water is insignificant when compared to the effect their presence has on reproducibility. Reproducibility is a difficult goal in the electrochemical deposition of any multi-metal films, and from

---

[†] Work in this chapter has formed the basis of two papers: ref. [21,22].

these investigations it has been found that the effects of the water and oxygen may be small compared to other effects but are more difficult to control. During the studies concerning film growth it was necessary to spend several days ensuring the purity of the dry box atmosphere, and if this important aspect was neglected then the standard of reproducibility would decrease. Once a stringent experimental procedure was determined it was possible to reproducibly deposit a variety of films with different compositions.

References

1   M. K. Wu, J. R. Ashburn, C. J. Torng, P. H. Hor, R. L. Meng, L. Gao, Z. J. Huang, Y. Q. Wang, and C. W. Chu, *Phys. Rev. Lett.*, 58 (1987) 905.

2   D. J. Zurawski, P. J. Kulesza, and A. Wieckowski, *J. Electrochem. Soc.*, 136 (1988) 1607.

3   M. Maxfield, H. Eckhardt, Z. Iqbal, F. Reidinger, and R. H. Baughman, *Appl. Phys. Lett.*, 54 (1989) 1932.

4   S. H. Pawar and H. A. Mujawar, *Mat. Res. Bull.*, 25 (1990) 1443.

5   S. H. Pawar and M. H. Pendse, *Mat. Res. Bull.*, 26 (1991) 641.

6   A. Weston, S. Lalvani, and N. Ali, *J. Mater. Sci.*, 2 (1991) 129.

7   A. Weston, S. Lalvani, F. Willis, and N. Ali, *J. Alloys and Compounds*, 181 (1992) 233.

8   P. Slezak and A. Wieckowski, *J. Electrochem. Soc.*, 138 (1991) 1038.

9   R. N. Bhattacharya, R. Noufi, L. L. Roybal, R. K. Ahrenkiel, P. Parilla, A. Mason, and D. Albin, *Science and Technology of Thin Film Superconductors 2*, Plenum Press, New York, 1990, pp. 243-250.

10   R. N. Bhattacharya, R. Noufi, L. L. Roybal, and R. K. Ahrenkiel, *J. Electrochem. Soc.*, 138 (1991) 1643.

11   R. N. Bhattacharya, P. A. Parilla, A. Mason, L. L. Roybal, R. K. Ahrenkiel, R. Noufi, R. P. Hellmer, J. F. Kwak, and D. S. Ginley, *J. Mater. Res.*, 6 (1991) 1389.

12   R. N. Bhattacharya, P. A. Parilla, R. Noufi, P. Arendt, and N. Elliot, *J. Electrochem. Soc.*, 139 (1992) 67.

13. R. N. Bhattacharya, P. A. Parilla, and R. D. Blaugher, *Physica C*, 211 (1993) 475.
14. R. N. Bhattacharya and R. D. Blaugher, *Physica C*, 225 (1994) 269.
15. R. N. Bhattacharya, A. Duda, D. S. Ginley, J. A. Deluca, Z. F. Ren, A. A. Wang, and J. H. Wang, *Physica C*, 229 (1994) 145.
16. J. M. Phillips, *Appl. Phys. Rev.*, 79 (1996) 1829.
17. A. Mogro-Campero, P. J. Bednarczyk, J. E. Tkaczyk, and J. A. Deluca, *Physica C*, 247 (1995) 239.
18. T. J. Mason, *Chemistry & Industry*, (1993) 47.
19. S. Osawa, M. Ito, K. Tanaka, and J. Kuwano, *Syn. Metals*, 18 (1987) 145.
20. A. Casaie and R. S. Porter, *Polymer Stress Reaction*, 2 (1979) 145.
21. K. A. Richardson, D. W. M. Arrigan, P. A. J. de Groot, P. C. Lanchester, and P. N. Bartlett, *Electrochimica Acta*, 41 (1996) 1629.
22. K. A. Richardson, P. A. J. de Groot, P. C. Lanchester, P. R. Birkin, and P. N. Bartlett, *J. Electroanal. Chem.*, 420 (1996) 21.

Chapter 5: Experimental Considerations for the Electrodeposition of....

Chapter 6:

Electrodeposition of Thallium-Based Superconductor Precursor Films

6 Electrodeposition of Thallium-Based Superconductor Precursor Films

6.1 Introduction

The previous chapter developed a reproducible experimental procedure through gaining control over the process environment and the solution composition. This chapter describes experiments in which this procedure was used to manufacture other superconductor precursor films, in particular, thallium-based films. By following the same procedure used in the synthesis of Bi-Sr-Ca-Cu films, precursor films of Ba-Ca-Cu, Hg-Ba-Ca-Cu, Tl-Ba-Ca-Cu, and Tl-Pb-Sr-Ca-Cu were produced. In most cases the deposition potential was either constant or pulsed. The reason for pulsing the applied potential, say between -3.5 V and -1.5 V at 0.05 Hz, is that during the -3.5 V pulse the metals are all deposited simultaneously and during the -1.5 V pulse only the nobler metals such as Tl and Cu are deposited maintaining the film conductivity, assisting in further deposition. This technique is thought to promote quality growth by limiting dendritic growth thereby producing denser films. As was shown in Chapter 4, the electrodeposition of Sr, Ca, and Ba was difficult to achieve because of the passive nature of the film formed. By depositing a highly conductive layer, after the co-deposition step, the inclusion of these passive components was made easier.

Details of the deposition bath composition, the resulting current vs. time profile, film morphology, and reproducibility for the deposited films are included. Also, the difficulty in growing films of Tl-Pb-Sr-Ca-Cu, because of the number of different components, was studied further by investigating the possibility of synthesising the films via a two-stage deposition process. Firstly a layer of Tl-Pb or Tl-Pb-Cu was laid down to be followed by a layer of Sr-Ca-Cu.

Finally, further work concerning the electrochemical considerations of the deposition of superconductor precursor materials is proposed - including multi stage approaches and ion complexing.

6.2 Experimental Procedure

The experimental procedure was similar for the electrodeposition of all the types of superconductor precursor films investigated. This procedure is described in this section and was applied to the production of the above listed precursor films.

All electrochemical experiments and preparation of deposition solutions were carried out in an Ar filled dry box to minimise contamination from water and oxygen. Dimethylsulfoxide (DMSO - Aldrich, 99.9 %) was used as-received but was opened in the dry box to avoid contamination from water. Nitrate salts of the relevant metallic components were stored over phosphorous pentoxide (P_2O_5) for approximately one month before use to ensure minimum water content. The composition of the deposition solution was adjusted in order to give the required stoichiometry in the precursor films. The substrates used for deposition of the precursor films were Ag foils (5.0 × 25.0 × 0.125 mm - Advent Research Materials, 99.7 %). They were polished with 0.3 μm alumina and cleaned with cotton wool and distilled water before use. Silicone 738 Adhesive/ Sealant (Dow Corning) was applied to mask the Ag substrates so that deposition only occurred on one side. All electrochemical deposition experiments were carried out in an undivided three-electrode cell with the reference, working, and counter electrodes suspended vertically. The relative positions of the three electrodes was held fixed from film to film, with a counter/working separation of ~ 1.0 cm. The reference electrode was a length of Ag wire (~ 80.0 mm), and the counter electrode was Pt foil (25.0 × 10.0 mm) covered with Pt gauze (Aldrich, 100 mesh 99.9 %). An EG&G model 273A potentiostat was employed to control the applied potential to the cell and monitor the current. The electrochemical cell was contained in a dri-block® DB-2A (Techne) heater within the dry box and the temperature maintained at 29.5 °C ± 0.1 °C. A fresh deposition solution was used in the synthesis of each film. After deposition the films were dried at 200 °C for 10 min.

The composition of the as-deposited films was determined using energy dispersive spectroscopy (EDS). The morphology of the as-deposited films was examined using a Jeol 2010 scanning electron microscope (SEM).

6.3 Electrodeposition of Ba-Ca-Cu Films

Films of Ba-Ca-Cu were produced with the correct 223 stoichiometry in order that Tl could be added during the post-deposition anneal programme rather than the electrodeposition stage. The films were produced by the application of a pulsed potential to the working electrode to cause deposition. The solution composition was adjusted in order to yield the desired film stoichiometry of Ba:Ca:Cu = 2:2:3 as determined by EDS. The bath contained 260.0 mM $Ba(NO_3)_2$, 130.0 mM $Ca(NO_3)_2H_2O$, and 46.0 mM $Cu(NO_3)_2H_2O$ dissolved in DMSO. The pulsed potential of -3.25 V, 10 s and -1.0 V, 10 s was applied for 1800 s at an ambient temperature of 30 °C. During this period a total charge of ~ 2140 mC was transferred. The current versus time profile for the deposition process is shown in figure 6.1. The envelope of the profile shows a peak current density of 13.0 mA cm^{-2} at t = 0. The subsequent fall, followed by a short rise in the current density can be attributed to the formation of nucleation centres on the working electrode followed by growth. From figure 6.2 it can bee seen that the initial occurrence of nucleation centres causes an increase in the area of the deposition surface. As the deposition progresses, and the nucleation sites grow to such an extent that neighbouring centres begin to meet and combine, the area of the layer then decreases, and then stabilises, causing the deposition rate to reach a point after which it is reasonably constant.

From the values of the redox potentials of the constituent elements determined in chapter 4 it was apparent that during the -3.25 V step all the constituents were co-deposited whereas during the -1.0 V step only $Cu^{(0)}$ was deposited with possibly some removal of loosely adhered Ba and Ca. From figure 6.1 it can be seen that during the -1.0 V step the current was anodic indicating that oxidation *was* occurring, and therefore material removal, this being combined with the deposition of copper metal. The total charge transferred for each of the -1.0 V pulses appeared to increase as time passed. This indicated that either the amount of copper deposited during the pulse decreased *or* the amount of material removed during the pulse increased (indicating dendritic growth because of an increasing surface area), *or* a combination of the two.

The large spikes present in the current versus time profile of figure 6.1 could be mistaken for the charging current due to the changing value of the double layer capacity when the applied potential was stepped. This can be represented by $I_{capacity} = (\Delta E/R)\exp(-t/RC)$ where R is the solution resistance (Ω), C is the double layer capacity (F), and ΔE is the change in the applied potential. When timescales involved were considered, though, the spikes were determined as being mainly faradaic - the charging current normally falls rapidly to zero in less than 50 µs and can be neglected for longer times.

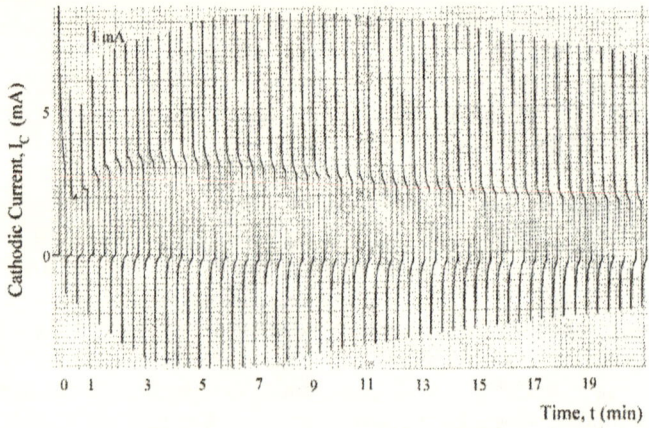

Figure 6.1 The current versus time profile for an electrodeposited Ba-Ca-Cu film synthesised using a pulsed-potential technique. The film was deposited from a solution comprised of 260.0 mM $Ba(NO_3)_2$, 130.0 mM $Ca(NO_3)_2H_2O$, and 46.0 mM $Cu(NO_3)_2H_2O$ dissolved in DMSO. The deposition was performed in a dry and the temperature maintained at 30 °C. A pulsed potential of -3.5 V, 10 s and -1.0 V, 10 s was applied for 1800 s.

Surface area increases as nucleation site grows

As the nucleation sites evolve they eventually begin to touch each other and the surface area therefore begins to decrease until the whole substrate is covered and the area stablises.

Figure 6.2 The affect of nucleation site evolution upon deposit surface area.

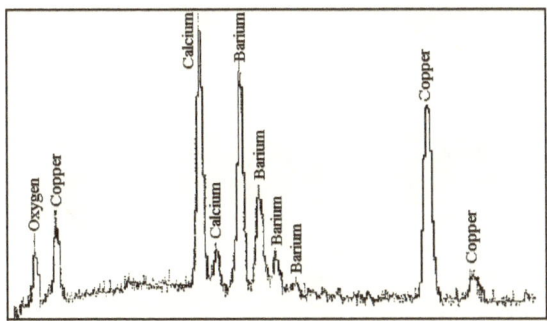

Figure 6.3 EDS spectrum for an as-deposited Ba-Ca-Cu film with the correct 223 stoichiometry.

EDS analysis of the films (figure 6.3) showed that they displayed a stoichiometry close to the desired 2:2:3 ratio, on average being Ba:Ca:Cu = 1.7:2.1:3.0. Moreover, the variation in the stoichiometry from film to film was extremely good, being no more than 10 % as determined by EDS. Figure 6.4 reflects this reproducibility by displaying the stoichiometry of several films. For large-scale production of these films to take place then the reproducibility must be maintained at an acceptable level. Finally, SEM analysis of the films shows that the film morphology comprises a homogeneous layer with clumps of growth on top (figure 6.5). The clumps may be the result of dendritic growth.

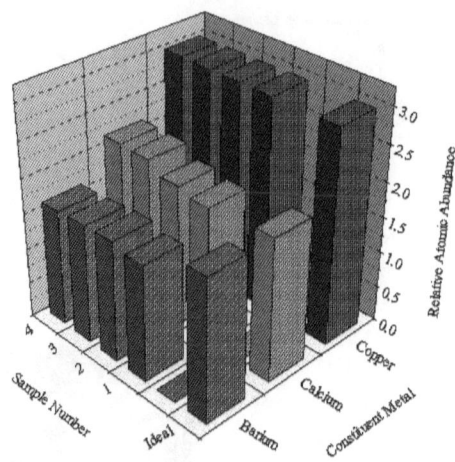

Figure 6.4 Reproducibility in the film stoichiometry of electrodeposited Ba-Ca-Cu films. The films were deposited from a solution comprised of 260.0 mM $Ba(NO_3)_2$, 130.0 mM $Ca(NO_3)_2H_2O$, and 46.0 mM $Cu(NO_3)_2H_2O$ dissolved in DMSO. The deposition was performed in a dry and the temperature maintained at 30 °C. A pulsed potential of -3.5 V, 10 s and -1.0 V, 10 s was applied for 1800 s. The solution was renewed between each deposition.

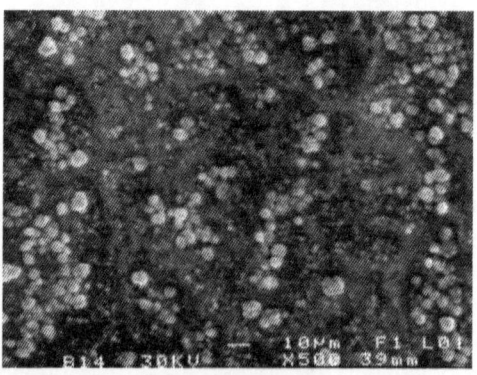

Figure 6.5 A SEM micrograph of an as-deposited Ba-Ca-Cu film. The film was deposited from a solution comprised of 260.0 mM $Ba(NO_3)_2$, 130.0 mM $Ca(NO_3)_2H_2O$, and 46.0 mM $Cu(NO_3)_2H_2O$ dissolved in DMSO. The deposition was performed in a dry and the temperature maintained at 30 °C. A pulsed potential of -3.5 V, 10 s and -1.0 V, 10 s was applied for 1800 s. (magnification: × 500).

6.4 Electrodeposition of Tl-Ba-Ca-Cu Films

Again, the deposition bath solution was systematically adjusted to yield the desired film stoichiometry of Tl:Ba:Ca:Cu = 2:2:2:3. A bath comprising of 5.6 mM TlNO$_3$, 500.0 mM Ba(NO$_3$)$_2$, 130.0 mM Ca(NO$_3$)$_2$H$_2$O, and 12.0 mM Cu(NO$_3$)$_2$H$_2$O, dissolved in DMSO was found to yield a stoichiometry close to the desired 2:2:2:3 ratio. An over abundance of thallium was included to compensate for the inevitable losses that would result from heat post-deposition treatment. A pulsed potential was applied across the cell to cause deposition onto individual Ag substrates. The applied potential was a pulse at -3.25 V and another at -1.0 V (both versus a Ag psuedo reference) at a frequency of 0.05 Hz and was applied for 1 hour. During this time a total charge of ~ 880.0 mC was transferred for films with an area of ~ 0.35 cm^2. The current versus time profile is depicted in figure 6.6.

The initial drop in the current vs. time envelope was due to the poor conductivity of the deposited film compared to the Ag substrate - the predicted film resistivity (assuming no phase formation) being ~ 10 µΩ cm [1] compared to 1.6 µΩ cm for silver at 20 °C. Note that the subsequent increase followed by a decrease in the profile that was observed in the Ba-Ca-Cu films was not observed for the deposition current vs. time profile for the Tl-Ba-Ca-Cu films. This may be because the nucleation occurred more rapidly in these films and so the effect was 'drowned out' by the pulsing effect.

From figure 6.6 it can also be seen that the current during the -1.0 V pulse was small. This was the result of a lower growth rate compared to the growth of the Ba-Ca-Cu films. The slower growth encouraged a more even film development and so less material was removed during the -1.0 V pulse. The anodic current due to material removal nearly equals the cathodic current due to copper *and* thallium deposition and hence a small net current was observed.

The resulting films composed typically of Tl:Ba:Ca:Cu = 2.0:1.7:1.8:3.0; close to the desired content, as determined by EDS. Figure 6.7 shows the EDS spectra obtained for such films. Notice that peaks derived from the Ag substrate are apparent. The films also appeared to contain significant amounts of oxygen, but it was not possible to tell whether the oxygen entered the film as part of the deposition process or absorbed from the atmosphere after removal from the dry box. The thickness of the films, as determined by SEM was found to be ~ 3 μm compared to a predicted thickness of ~ 1 μm derived from the total charge transferred (880 mC). This yielded a void percentage of $([3-1]/3) \times 100$ = 67 %. This is significantly lower than the value of 90 % obtained for Bi-Sr-Ca-Cu films, but not as good as the 45 % obtained for ultrasonically electrodeposited films. The probable cause for the difference between pulsed and constant potential deposition is that the pulsed potential approach encourages denser film growth because of the removal of poorly adhered material during the -1.0 V step, limiting dendrite formation.

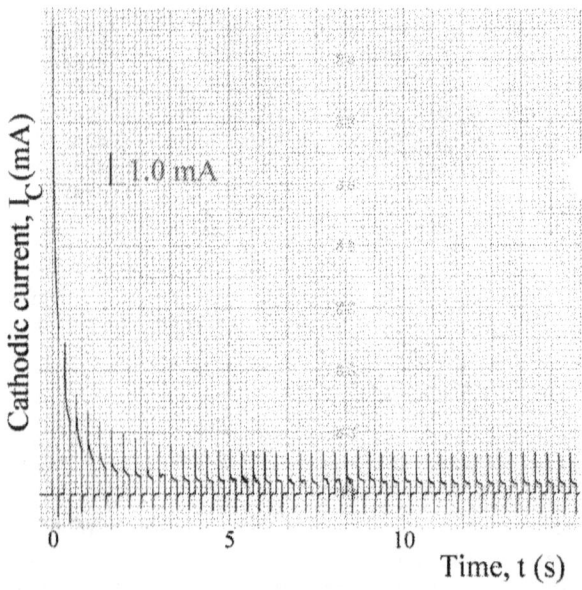

Figure 6.6 Current versus time for an electrodeposited Tl-Ba-Ca-Cu film produced using a pulsed-potential technique. The film was deposited from a solution comprised of 5.6 mM $TlNO_3$, 500.0 mM $Ba(NO_3)_2$, 130.0 mM $Ca(NO_3)_2H_2O$, and 12.0 mM $Cu(NO_3)_2H_2O$ dissolved in DMSO. The deposition was performed in a dry and the temperature maintained at 30 °C. A pulsed potential of -3.5 V, 10 s and -1.0 V, 10 s was applied for 3600 s.

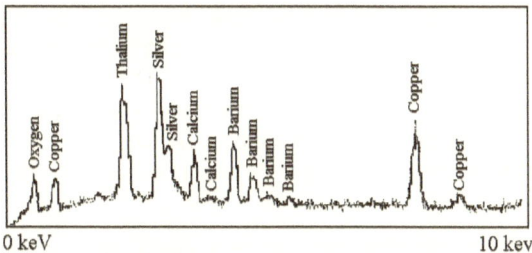

Figure 6.7 EDS *spectra of Tl-Ba-Ca-Cu films with a stoichiometric composition of Tl:Ba:Ca:Cu = 2.0:1.7:1.8:3.0.*

Figure 6.8 is an SEM micrograph of a Tl-Ba-Ca-Cu film which indicates the good large-scale homogeneity of the films produced.

Figure 6.8 An SEM *micrograph of an electrodeposited Tl-Ba-Ca-Cu film with a stoichiometric composition close to the desired 2:2:2:3 ratio. The film was deposited from a solution comprised of 5.6 mM TlNO$_3$, 500.0 mM Ba(NO$_3$)$_2$, 130.0 mM Ca(NO$_3$)$_2$H$_2$O, and 12.0 mM Cu(NO$_3$)$_2$H$_2$O dissolved in* DMSO. *The deposition was performed in a dry and the temperature maintained at 30 °C. A pulsed potential of -3.5 V, 10 s and -1.0 V, 10 s was applied for 3600 s. (magnification: × 1000).*

6.5 Electrodeposition of Tl-Pb-Sr-Ca-Cu Films

The principal difficulty in the electrodeposition of Tl-Pb-Sr-Ca-Cu (TPSCC) films over the synthesis of other films presented herein, was the greater number of metallic cations that required to be co-deposited. The difficulty being in obtaining the correct stoichiometry. The attraction to production of these films is that when reacted through an appropriate heat treatment programme to form the 1223 phase, the critical current, and the critical current applied field dependence is superior to Tl-Ba-Ca-Cu films [2]. This improvement in transport properties is attributed to the Pb doping. This doping introduces pinning centres to the crystal lattice, reducing the movement of magnetic field lines (vortices) through the film.

Several electrochemical approaches have been attempted to synthesis TPSCC precursor films. Firstly, films have been produced using a constant potential technique, similar to the technique used in the deposition of Bi-Sr-Ca-Cu films (§ 5.4.2). Secondly, the films were produced by applying a pulsed potential, and finally, a two stage deposition process was investigated by depositing a Tl-Pb-Cu film at one potential and then a Sr-Ca-Cu film over deposited at a higher potential. Each one of these methods is discussed.

6.5.1 Constant Potential Deposition of TPSCC Films

A deposition bath was prepared containing 12.0 mM $TlNO_3$, 2.0 mM $Pb(NO_3)_2$, 200.0 mM $Sr(NO_3)_2$, 480.0 mM $Ca(NO_3)_2H_2O$, and 16.0 mM $Cu(NO_3)_2H_2O$ dissolved in DMSO. The systematic optimisation procedure of the deposition bath composition was achieved in the same way as the solution composition was determined for Bi-Sr-Ca-Cu films in chapter 5. A constant potential of -3.25 V (versus a Ag pseudo reference) was applied to the Ag substrate for 3600 s to cause co-deposition of the components. Figure 6.9 shows the I/t curve for the deposition process, indicating the familiar initial drop due to the reduced conductivity of the deposited films compared to the Ag substrate. During the deposition a charge of 1200 mC was transferred typically for films with an area of ~ 0.35 cm^2.

EDS analysis showed that the films had a composition close to the desired stoichiometry of Tl:Pb:Sr:Ca:Cu = 4.8:0.6:1.9:1.8:3.0. However, it was found to be difficult to control the calcium content sufficiently to gain satisfactory reproducibility using this approach, in spite of the fact that it caused no difficulties in the deposition of Bi-Sr-Ca-Cu precursor films. To solve this problem an experiment was performed whereby a TPSCC film was deposited in the same way as described above yielding a film stoichiometry close to the desired stoichiometry. The transfer chamber door of the dry box was then opened briefly (5 s) and then closed. Another film was deposited from a fresh solution which yielded a film with almost the same composition as before except that the calcium abundance had doubled. The reason for this can be attributed to that fact that the solution composition for the deposition of the TPSCC contains large amounts of $Ca(NO_3)_2H_2O$ (480.0 mM) which is extremely hygroscopic. This makes the solution composition more sensitive to the presence of water than the solution for BSCC (which contained relatively small amounts of $Ca(NO_3)_2H_2O$. Any leaks in the dry box or carelessness in minimising the oxygen and water content of the dry box atmosphere would result in a change of solution composition and hence film stoichiometry.

Good film homogeneity was achieved as seen can be observed from the SEM micrograph shown in figure 6.10.

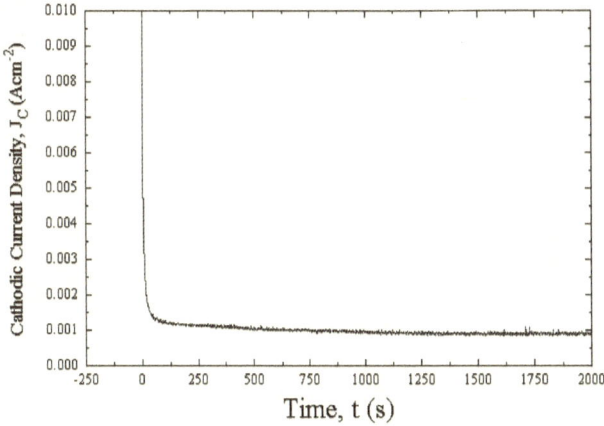

Figure 6.9 Current versus time for a TPSCC film deposited at -3.25 V for 3600 s. The film was deposited from a solution comprised of 12.0 mM $TINO_3$, 2.0 mM $Pb(NO_3)_2$, 200.0 mM $Sr(NO_3)_2$, 480.0 mM $Ca(NO_3)_2H_2O$, 16.0 mM $Cu(NO_3)_2H_2O$. The deposition was performed in a dry box and the temperature was maintained at 30 °C.

6:11

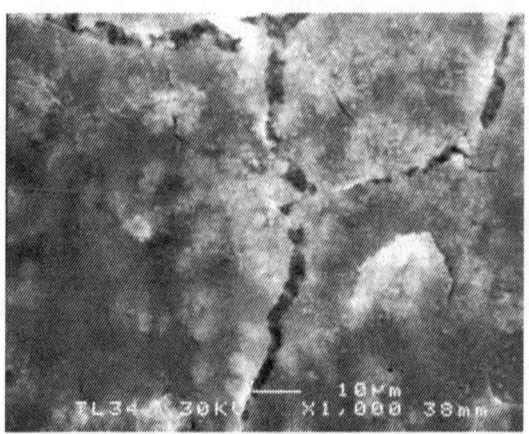

Figure 6.10 SEM micrograph of as-deposited TPSCC film produced using constant potential electrodeposition. The film was deposited from a solution comprised of 12.0 mM $TINO_3$, 2.0 mM $Pb(NO_3)_2$, 200.0 mM $Sr(NO_3)_2$, 480.0 mM $Ca(NO_3)_2H_2O$, 16.0 mM $Cu(NO_3)_2H_2O$. The deposition was performed in a dry box and the temperature was maintained at 30 °C. (magnification: × 1000).

6.5.2 Pulsed Potential Deposition of TPSCC Films

The constant potential deposition of TPSCC lead to a low deposition rate. The bath prepared above leads to the maximum rate for that system - when the bath salt concentrations were all doubled an increase of only 100.0 mC was achieved in the total charge transferred, a disproportional increase given the materials used. A pulsed potential deposition technique was once again employed in an effort to increase the deposition rate. The effect of using a pulsed potential was discussed in section 6.5.

A deposition bath comprising 12.0 mM $TINO_3$, 2.0 mM $Pb(NO_3)_2$, 225.0 mM $Sr(NO_3)_2$, 450.0 mM $Ca(NO_3)_2H_2O$, and 16.0 mM $Cu(NO_3)_2H_2O$ dissolved in DMSO was prepared, very similar to the bath prepared for constant potential deposition. A pulsed potential was applied to the working electrode: the potential being -3.5 V, 10 s and -1.5 V, 10 s. This cycle was repeated for 3600 s. The current versus time profile produced using the pulsed approached in shown in figure 6.11. The profile envelope is similar to that for the constant potential approach, with the initial (maximum) current density of 21.6 mA cm^{-2}, and the current density levelling at around 2.3 mA cm^{-2} after approximately 7 minutes.

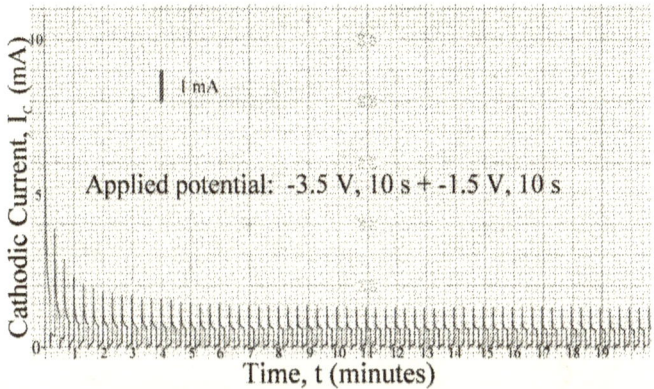

Figure 6.11 Current versus time profile for a pulsed potential produced TPSCC film. The film was deposited from a solution comprised of 12.0 mM TlNO$_3$, 2.0 mM Pb(NO$_3$)$_2$, 225.0 mM Sr(NO$_3$)$_2$, 450.0 mM Ca(NO$_3$)$_2$H$_2$O, 16.0 mM Cu(NO$_3$)$_2$H$_2$O. The deposition was performed in a dry box and the temperature was maintained at 30 °C. The potential was pulsed from -3.5 V, 10s to -1.5 V, 10 s for 3600 s.

The charge transferred for each film during the 3600 s deposition period was 3.6 ± 0.2 C cm^{-2}. This is similar to the charge transferred during the deposition procedure for a constant potential process (1450 mC for a film area of ~ 0.4 cm^2). Note also from the figure that during the -1.5 V step the current was cathodic indicating that the deposition of Tl, Pb, and Cu was more dominant than the removal of poorly adhered Sr and/or Ca. It may be possible to increase the rate by adjusting the magnitude of the potential steps and/or the pulse duration.

EDS analysis (figure 6.12) showed that the films had the desired stoichiometry of Tl:Pb:Sr:Ca:Cu = 4.6:0.5:1.6:2.1:3.0 - the over abundance of Tl being included to compensate for Tl loss expected during the post deposition heat treatment. Yet, again the reproducibility of the films' stoichiometry was found to be of a high standard, yielding a variation of no more than 10 % as determined by EDS analysis (figure 6.13).

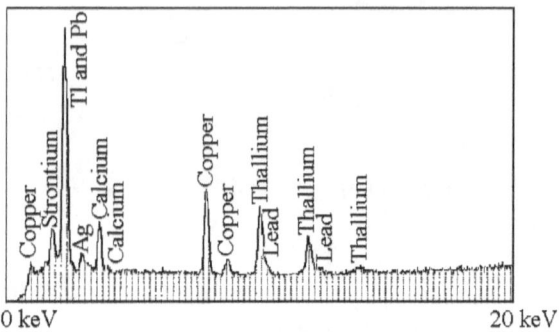

Figure 6.12 EDS spectrum of as-deposited TPSCC film produced using a pulsed potential technique. The spectrum indicates a film composition of Tl:Pb:Sr:Ca:Cu = 4.6:0.5:1.6:2.3:3.0.

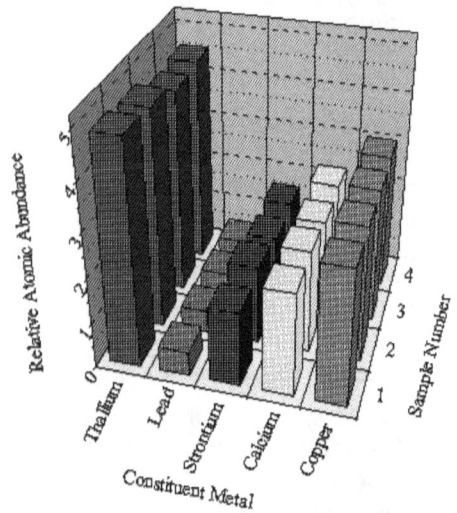

Figure 6.13 The stoichiometric reproducibility for as-deposited TPSCC films. The film was deposited from a solution comprised of 12.0 mM $TlNO_3$, 2.0 mM $Pb(NO_3)_2$, 225.0 mM $Sr(NO_3)_2$, 450.0 mM $Ca(NO_3)_2H_2O$, 16.0 mM $Cu(NO_3)_2H_2O$. The deposition was performed in a dry box and the temperature was maintained at 30 °C. The potential was pulsed from -3.5 V, 10s to -1.5 V, 10 s for 3600 s. The deposition solution was changed for each film.

Figure 6.14 shows the films' morphology produced using the pulsed potential approach. The morphology was homogeneous over the area of the film and the largest features were of the order of 10 μm in size.

Figure 6.14 SEM micrograph of an as-deposited TPSCC film produced using a pulsed potential method. The film was deposited from a solution comprised of 12.0 mM $TlNO_3$, 2.0 mM $Pb(NO_3)_2$, 225.0 mM $Sr(NO_3)_2$, 450.0 mM $Ca(NO_3)_2H_2O$, 16.0 mM $Cu(NO_3)_2H_2O$. The deposition was performed in a dry box and the temperature was maintained at 30 °C. The potential was pulsed from -3.5 V, 10s to -1.5 V, 10 s for 3600 s. (magnification: × 500).

6.5.3 Towards a Two Stage Process for the Manufacture of TPSCC films

Because of the difficulties in growing very thick films of TPSCCO precursors the possibility of synthesising these films by a two stage electrochemical procedure was investigated. Firstly, a layer of Tl and Pb or Tl, Pb and Cu would be deposited followed by an overcoat of Sr, Ca and Cu. Attempts were made to produce Sr-Ca films but the passive nature of these materials restricted deposition to the synthesis of only thin films (~ 0.1 μm) and so Cu had to be included to ensure that the deposited layer would not hinder further deposition. A possible advantage of this approach would be that all the constituents, except Cu, would be deposited close to their respective redox potentials and hence dendritic growth (and therefore high void percentages) may be avoided. Moreover, as the Tl and Pb would be under a thick film of Sr-Ca-Cu mix then, during the post-deposition heat treatment, Tl and Pb might not evaporate so quickly as they would have to diffuse through the upper layer. A distinct disadvantage, however, would be that, as the films' constituents would no longer be well-mixed on an atomic scale, the post-production heat treatment programme might not result in monophasic films. In this section the deposition conditions for the synthesis of Tl-Pb, Tl-Pb-Cu, and Sr-Ca-Cu films are presented.

Table 6.1 shows the deposition bath composition used for the three types of films produced. Both the Tl-Pb and the Tl-Pb-Cu films were deposited by applying a constant potential of -1.30 V for 1800 s, at 30 °C, whereas the Sr-Ca-Cu films were produced via the application of -3.25 V for 3600 s and also at 30 °C. Figure 6.15 shows the current versus time profiles for the different films. The initial current density for Tl-Pb, Tl-Pb-Cu, and Sr-Ca-Cu are 2.7 mA cm^{-2}, 13.0 mA cm^{-2}, and 20.0 mA cm^{-2}, respectively. The current versus time profiles of the Tl-Pb and Tl-Pb-Cu both fell to a minimum of 5.2 mA cm^{-2} and 1.8 mA cm^{-2}, respectively, and then rose to a current density of 2.2 mA cm^{-2} and 7.6 mA cm^{-2}, respectively, for the remaining duration of the deposition period. The general structure of both these current vs. time profiles is indicative of the deposition of highly conductive metallic layers. The current versus time profile for the synthesis of Sr-Ca-Cu films shows an rapid drop to a current density of 1.2 mA cm^{-2}, very similar to the deposition profile for a superconductor precursor formed via a single stage deposition process. During the deposition period the total charge transferred was 1360 mC, 4800 mC, and 1160 mC for deposited films of Tl-Pb, Tl-Pb-Cu, and Sr-Ca-Cu, respectively. From EDS analysis the compositions of the films were found to be Tl:Pb = 3.0:0.5, Tl:Pb:Cu = 2.5:0.6:1.0, and Sr:Ca:Cu = 2.0:2.0:2.0 which are ideal, in principle, for the application of a two-stage procedure.

Film Type	Bath Composition
Tl-Pb	$TlNO_3$ = 8.0 mM
	$Pb(NO_3)_2$ = 1.5 mM
Tl-Pb-Cu	$TlNO_3$ = 8.0 mM
	$Pb(NO_3)_2$ = 1.5 mM
	$Cu(NO_3)_2$ = 10.0 mM
Sr-Ca-Cu	$Sr(NO_3)_2$ = 189.0 mM
	$Ca(NO_3)_2H_2O$ = 330.0 mM
	$Cu(NO_3)_2H_2O$ = 24.3 mM

Table 6.1 Electrochemical bath compositions used for the synthesis of Tl-Pb, Tl-Pb-Cu, and Sr-Ca-Cu films.

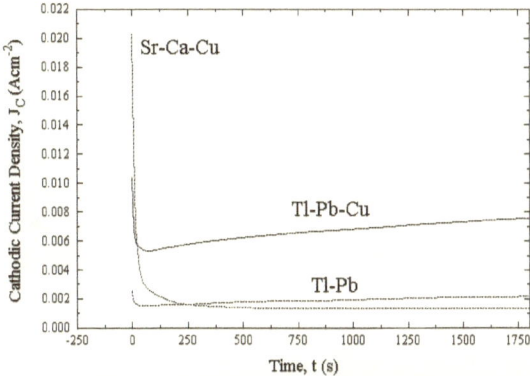

Figure 6.15 Current versus time profiles for electrodeposited films of Tl-Pb, Tl-Pb-Cu, and Sr-Ca-Cu. The composition of the solutions from which the films were deposited are shown in table 6.1. All films were deposited in a dry box and the temperature regulated at 30 °C. The Tl-Pb and the Tl-Pb-Cu films were deposited at a constant potential of -1.3 V for 1800 s, whereas the Sr-Ca-Cu films were deposited at a constant potential of 3.5 V for 3600 s.

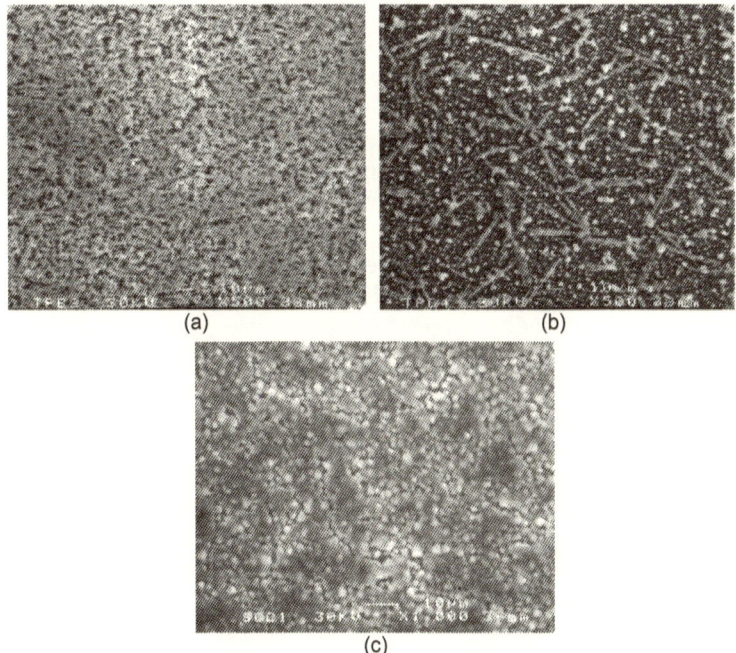

Figure 6.16 SEM micrographs of: (a) Tl-Pb (magnification: × 500), (b) Tl-Pb-Cu (magnification: × 500), and (c) Sr-Ca-Cu as-deposited (magnification: × 1000). See table 6.1 for deposition bath composition.

Figure 6.16a, b and c are SEM micrographs of the three types of film produced for the two-stage deposition process. All three display very different morphologies.

6.6 Electrodeposition of Hg-Ba-Ca-Cu Films

A set of experiments were performed in an attempt to include Hg into the Ba-Ca-Cu precursor films. The problem with Hg is similar to that of Cu in that it has a relatively low redox potential. The electrochemical bath was not adjusted to obtain a particular stoichiometry, the aim being only to observe the incorporation of Hg into the electrodeposited films.

A deposition bath was prepared comprising 4.2 mM $HgNO_3$, 13.0 mM $Ba(NO_3)_2$, 11.5 mM $Ca(NO_3)_2H_2O$, and 3.0 mM $Cu(NO_3)_2H_2O$ dissolved in DMSO. A pulsed potential was applied of: -4.0 V, 10 s, and -1.0 V, 10 s. This was cycled for 25 minutes to achieve deposition of the films. The fabrication was performed at 18 °C and stirring of the solution was also included. The difference in applied potential from most other deposition reported herein was due to the difference in deposition temperature.

The deposition process of the Hg-Ba-Ca-Cu films caused 660.0 mC of current to be transferred during the 25 minute deposition time - the films having an area of ~ 0.5 cm^2. EDS analysis of the films afforded film compositions of Hg:Ba:Ca:Cu = 20.0:2.8:3.9:3.0. This, apart from having slightly incorrect Ba and Ca contents, shows a large over abundance of Hg. This being due to the relatively small redox potential of Hg^+ (-0.05 V versus Ag psuedo reference) as well as the relative concentration of Hg in the deposition solution. This observed over abundance may be desirable. Hg is very volatile, not unlike Tl, and so during any heat treatment it would be expected that significant amounts of Hg would evaporate, thus, a relatively large Hg content may be necessary for successful conversion to the superconducting phase. In fact, the synthesis of mercury-based superconducting powder is usually performed under high pressure to avoid large amounts of mercury loss. The SEM micrograph represented in figure 6.17 shows the large amount of Hg present as globules of Hg metal (as determined from EDS).

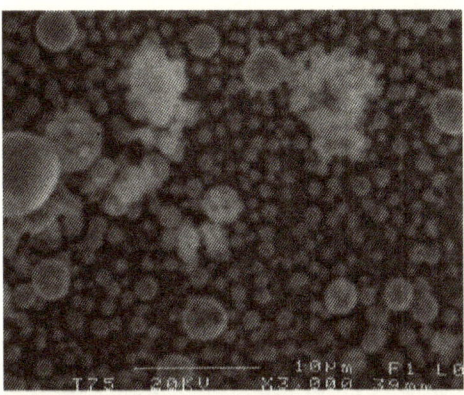

Figure 6.17 A SEM micrograph of an electrodeposited Hg-Ba-Ca-Cu film clearly showing the excess of Hg metal. (magnification: ×3000). The film was deposited from a solution comprising 4.2 mM $HgNO_3$, 13.0 mM $Ba(NO_3)_2$, 11.5 mM $Ca(NO_3)_2H_2O$, and 3.0 mM $Cu(NO_3)_2H_2O$ dissolved in DMSO. The deposition was performed in a dry box and the temperature maintained at 18 °C. The potential was pulsed at -4.0 V, 10 s and -1.0 V, 10 s for 25 minutes. (magnification: ×3000)

6.7 Discussion and Further Work

In this chapter the electrodeposition of Ba-Ca-Cu, Tl-Ba-Ca-Cu, Tl-Pb-Sr-Ca-Cu, and Hg-Ba-Ca-Cu superconductor precursor films has been presented. It was found that all the types of films could be deposited routinely with a high standard of reproducibility, similar to that of the Bi-Sr-Ca-Cu films presented in chapter 5. The standard of the reproducibility was heavily dependent on the control of the different aspects of the process, described in detail in chapter 5. Through experience, one of the most difficult parameters to control was found to be the water content of the deposition solution, and the utmost care must be taken to keep the water level to a minimum - only then can this standard of reproducibility be obtained. The reproducibility in the film stoichiometries of the thallium-based films was found to be harder to maintain than the bismuth-based films. This was possibly because of the increased sensitivity of the thallium-based deposition solutions which arose from the use of the large amounts of hygroscopic ionic salts dissolved within, e.g. $Ca(NO_3)_2H_2O$.

To overcome this severe water problem it may be possible to produce complexed solutions of the metallic salts that do not include water. For example, the water/metal complex, may be converted into a DMSO/metal complex. Work performed by Pawar et al. (reference [4] in chapter 5) already investigated the effect of complexing a Bi-Sr-Ca-Cu solution with sodium citrate, EDTA, sodium nitrate, and tartaric acid. In this case EDTA and sodium citrate were found to be suitable complexing agents in an acetone bath producing more uniform and crystalline films compared to the uncomplexed bath. More recently Ondoño-Castillo et al. [3] complexed copper in a Y-Ba-Cu bath with cyanide. The effect of this complex was to move the redox potential of the Cu^{2+} closer to that of the yttrium and the barium. The resulting films were ~ 100 µm thick but the superconducting properties were surprisingly poor. This may be due to a non-optimised heat treatment programme rather than low precursor film quality. Research is currently (as of 1996) being performed by Dr. G. Goodlet, a post-doctoral member of the Southampton group, in order to learn more about the affect of ion complexing upon the deposition procedure. The aim is to move the redox potentials of the relatively easily reduced cations closer to the more difficulty reduced cations. It is important to remember the lessons learnt from using pulsed deposition. Pulsed deposition provides a route to synthesising high quality precursors with void percentages of ~ 67 % compared with the 92 % obtained by constant potential deposition. This means that thicker *quality* films could be fabricated using the pulsed applied potential approach as opposed to the constant applied potential approach. Therefore, the aim would not be to encourage copper, lead, thallium, mercury, and bismuth to reduce at a similar potential to that of strontium, calcium, and barium because this will mean that a constant potential will have to be applied and hence losing part of the benefit gained through using a pulsed approach. The aim will be, therefore, to move to redox potentials of copper, lead, bismuth, thallium, and mercury to a redox potential slightly lower (by, say, ~ 500 mV) than the redox potentials of strontium, barium, and calcium in order that the full benefits of pulsed deposition may be incorporated into the procedure. This work could be coupled with further investigations concerning the inclusion of an ultrasonic radiation source into the process, having already established the significant benefits of this addition in chapter 5 and previously published results [4-8].

The environmental reactivity of these materials deserves comment. It is already known that some of the HTSC phases react rapidly with water, CO_2, CO, and acids [9,10]. The prevention of the degrading properties of the films should be provided. Several publications [11-15] have successfully electrodeposited a protective layer of copper into the superconducting films to prevent such degradation of the films. Again, Dr. G. Goodlet is currently investigating this area. In particular research into depositing layers of silver onto the films before and/or after the heat treatment procedure. The influence upon film properties would also need to be quantified.

References

1. Calculated by averaging the resistivities of Tl, Pb, Sr, Ca, and Cu metal obtained from *CRC Handbook of Chemistry and Physics*, 1995-1996.
2. T. Kamo, T. Doi, A. Soeta, T. Yuasa, N. Inoue, K. Aihara, and S. - P. Matsuda, *Appl. Phys. Lett.*, 59 (1991) 3186.
3. S. Ondoño-Castillo, A. Fuertes, F. Pérez, P. Gómez-Romero, and N. Casañ Pastor, *Chem. Mater.*, 7 (1995) 771.
4. S. R. Rich, Proc. Am. Electroplaters' Soc., 42 (1955) 137.
5. K. S. Suslick, J. J. Gawienowski, P. F. Schubert, and H. H. Wang, *Ultrasonics*, January (1984) 33.
6. K. S. Suslick, *Science*, 247 (1990) 1439.
7. T. J. Mason, *Chemistry & Industry*, January (1993) 47.
8. H. Zhang and L. A. Coury, Jr., *Anal. Chem.*, 65 (1993) 1552.
9. J. -P. Zhou and J. T. McDevitt, *Chem. Mater.*, 4 (1992) 953.
10. J. -P. Zhou, D. R. Riley, A. Manthiram, M. Arendt, M. Schmerling, and J. T. McDevitt, *Appl. Phys. Lett.*, 63 (1993) 548.
11. N. A. Fleischer and J. Manassen, *J. Electrochem. Soc.*, 135 (1988) 3174.
12. J. M. Rosamilia and B. Miller, *J. Electrochem. Soc.*, 136 (1989) 1053.
13. Y. S. Chang, S. M. Ma, F. L. Yang, and C. S. Li, *Maters. Chem. Phys.*, 28 (1991) 121.
14. F. L. Yang, Y. S. Chang, C. S. Li, S. M. Ma, Y. T. Huang, and W. H. Lee, *J. Maters. Sci.*, 27 (1992) 5739.
15. S. M. Ma, Y. S. Chang, F. L. Yang, C. S. Li, Y. T. Huang, and W. H. Lee, *J. Electrochem. Soc.*, 139 (1992) 1951.

Chapter 7:

Superconducting Properties of Electrodeposited Films

7 Superconducting Properties of Electrodeposited Films

7.1 Introduction

The electrochemical section of this thesis has thus far concentrated mainly on the electrodeposition of superconductor precursor films. Because of difficulties in attaining high levels of reproducibility, the mechanisms that influence the deposition process were investigated in detail so that the insight developed would contribute to the development of a continuous manufacturing process. Films with the correct starting stoichiometry were prepared and analysed comprehensively. The success of routinely preparing these films, given the sensitivity of the process to many factors, was taken a step further. This additional step attempted to induce superconductivity in to the precursor films through appropriate heat treatment. In this chapter the film characteristics of post annealed Bi-Sr-Ca-Cu and Tl-Pb-Sr-Ca-Cu films are presented. The analysis concentrates on the details of film phase and purity, film texturing, superconducting transition temperature, and magnetic properties. XRD was employed to yield phase and texturing data. SQUID magnetometry, AC susceptibility, and VSM measurements were used to acquire data concerning the superconducting properties. These measurements by no means form a comprehensive study of the films produced, but represent initial measurements to determine basic superconducting properties.

7.2 Superconductivity in Bi-Sr-Ca-Cu Films

The fabrication of Bi-Sr-Ca-Cu films, described in chapter 5, produced porous films with stoichiometries close to the desired 2212 ratio. These precursors then had to be reacted to form the superconducting phase. The sintering was performed in a tube furnace with an air atmosphere. Extensive optimisation of the furnace temperature and sintering time was required in order that single phase, $Bi_2Sr_2CaCu_2O_{8+\delta}$ films were obtained. This systematic study determined the sintering temperature to be 810 °C with a sintering time of 6 hours. All samples were furnace cooled at a rate of ~ 3.0 °C min^{-1}. During sintering the samples were simply placed on top of a ceramic crucible. No special mounting procedure was used.

The as-deposited films were slightly brown in appearance with a porous morphology, with grouped 'islands' of deposit. The size of the 'islands' ranged from 1.0 µm upto 13.0 µm across. The sintered films, however, were very black and appeared matt to the naked eye. The morphology had evolved to appear denser, with a feature size of ~ 8.0 µm – which was assumed to be the superconducting grain size, D, for certain calculations. The change in morphology can be seem in SEM's of the films before an after the heat treatment (figure 7.1). Both before and after heat treatment the films seemed to be well adhered, and even after modest bending of the substrates the films remained fixed to the substrate indicating that the films were reasonably ductile.

By tilting the sample during SEM analysis it was possible to calculate the thickness of the heat treated films. For the Bi-Sr-Ca-Cu films the thickness was found to be ~ 23.0 µm. This was significantly thinner than the as-deposited films indicating that densification arose as a result of the heat treatment. Re-calculating the void percentage, a new value of ~ 80 % was obtained compared to ~ 92 % before heat treatment. This value is rather high, but might be reduced further by longer reaction times whereby increased diffusion would occur, and therefore movement, of the film constituents.

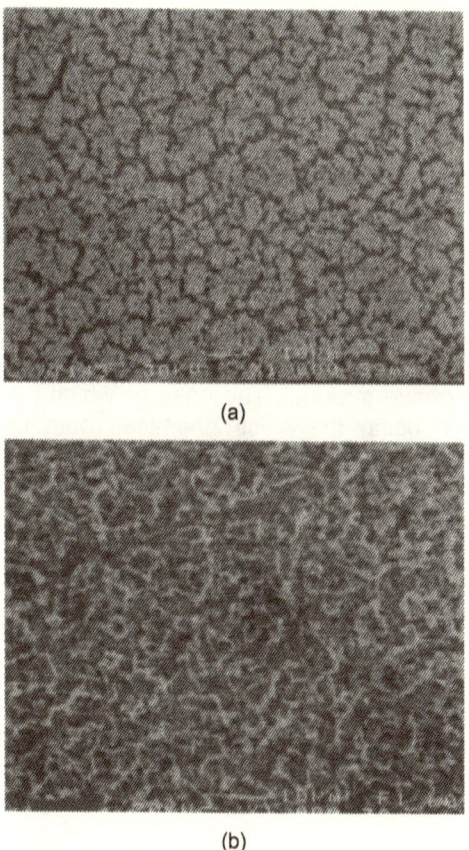

(a)

(b)

Figure 7.1 SEM micrographs of a (a) as-deposited Bi-Sr-Ca-Cu film, and (b) post-sintered Bi-Sr-Ca-Cu film. The films were deposited film a solution comprising 6.0 mM $Bi(NO_3)_3$, 6.0 mM $Sr(NO_3)_2$, 5.0 mM $Ca(NO_3)_2 H_2O$, and 7.0 mM $Cu(NO_3)_2 H_2O$ dissolved in DMSO. Deposition was performed in a dry box with the temperature maintained at 30 °C. Deposition was caused by applying a constant potential of -3.25 V across the cell for 1800 s. The as-deposited films were then dried at 200 °C for 10 minutes and then sintered at 810 °C in air for 6 hours followed by a furnace cool. (Magnification: × 1000)

The superconducting phase that formed the films was then determined. The predicted X-ray spectrum for a c-axis aligned (00l) Bi-2212 sample can be seen in figure 7.2. If this spectrum is compared to the spectra depicted in figure 7.3 of sintered Bi-Sr-Ca-Cu films then it is clearly apparent that the films obtained displayed a high level of texturing. Moreover, from the X-ray analysis the films were very nearly single phase

$Bi_2Sr_2CaCu_2O_{8+\delta}$ superconductors. The two spectra in figure 7.3 were taken from two different films deposited from different baths on separated days. From the spectra the level of reproducibility is clear, the films both having very similar structures. The slight variation in peak intensities was probably due to slight differences in the mounting position within the X-ray spectrometer. As the samples are relatively small, compared to the beam size, the position of the sample is important in obtaining consistent XRD measurements. Another favourable feature is the lack of impurity phases present in the samples. Using powder techniques in the production of bismuth-based samples (or for that matter any superconducting material) rarely yields single phase material. The ease by which the pure phase is synthesised using the electrochemical route is due to the indomitable mixing of the process - mixing on an atomic scale.

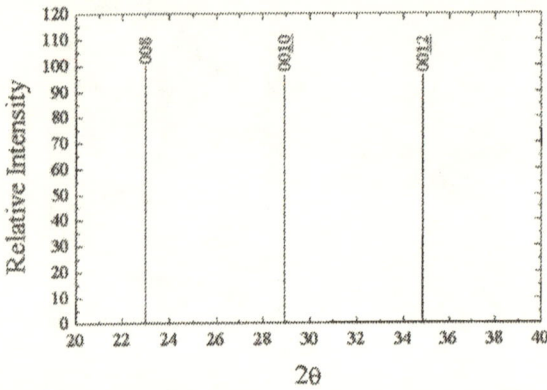

Figure 7.2 Theoretical spectrum for a c-axis aligned Bi-2212 film.

Also, to be noted is the presence of peaks other than the desired 00l lines. The existence of these lines is due to the mismatch between the Ag substrate and the superconductor lattice parameters. As mentioned previously, Ag was used because of its availability and low cost. If better c-axis alignment was required then other substrates, such as Ag coated MgO, should be used. Film texturing is required to achieve good transport properties (not quantified herein). This is because of the highly anisotropic nature of the present HTSC materials.

Chapter 7: Superconducting Properties of Electrodeposited Films

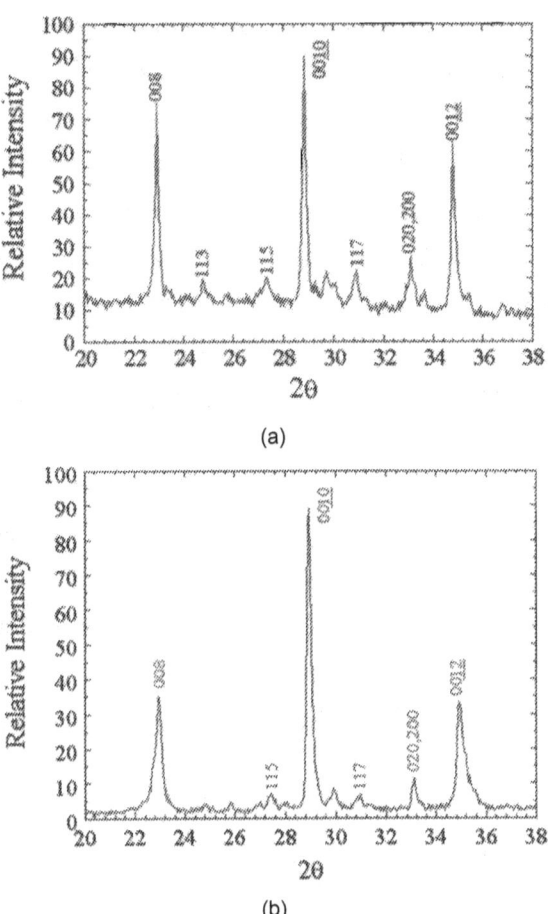

Figure 7.3 XRD spectra of two post-sintered $Bi_2Sr_2CaCu_2O_{8+\delta}$ films. Note the excellent c-axis alignment (texturing).

Next, the T_C of the films was quantified with SQUID magnetometry. The sensitivity of the SQUID was ideal considering the small amounts of material under examination. Figure 7.4 presents the results from these measurements of sample moment versus temperature. From the graph it is apparent that a clear T_C of ~ 87 K exists. This value exceeds that of previous attempts to produce $Bi_2Sr_2CaCu_2O_{8+\delta}$ via an electrochemical step [1], and is of comparable magnitude to those of powder samples synthesised via the 'shake and bake' approach [2].

In many heat treatment programmes, after the initial sintering step, an additional step is included. This step is normally a post reaction low-temperature anneal step, whereby the oxygen stoichiometry is optimised to achieve the best possible value, and hence improved superconducting properties. Inter-grain connectivity may sometimes be enhanced. Selected films were annealed in an O_2 atmosphere at 450 °C. This was continued for 2 - 5 hours and the samples were then furnace cooled as before. AC susceptibility, χ_{AC} measurements were then performed on small sections of the films to measure any changes in the T_C due to the anneal stage. Figure 7.5 shows a typical variation of the real and imaginary parts of χ_{AC} versus temperature. The T_C had increased by an impressive 10 - 12 K. This value of ~ 100 K is exceptionally high for this material and promising from the point of applications. The large improvement over previous attempts to produce bismuth-based films may be due to the differing experimental procedures used in the fabrication of the precursor films. This may indicate that the quality of the precursor films has an influence upon the intrinsic properties of the phases formed after heat treatment.

The shape of the transition in χ_{AC} measurements is not typical when compared to the crystal or pellet measurements. The transitions appeared to be broad (at very small field) and did not actually reach the point at which perfect diamagnetism was achieved. This anomaly is an inherent property of film samples, and can be explained by considering the magnetisation factor, N, of such samples - and is especially important when trying to understand susceptibility in films. In the equation for magnetic induction, $B = \mu_0(H + M)$, where H is the internal field, equal to the applied field, $H_{applied}$, corrected by the demagnetising field, H_{demag}. The source of the demagnetising field is taken to be magnetic poles on the surface of a magnetised sample. The magnetic susceptibility of superconductor films, measured in perpendicular fields, presents a paradox arising from perfect diamagnetism and a demagnetising factor that approaches unity. For perfectly shielding superconductors, $\chi = -1$, so $\chi_{external} = -(1 - N)^{-1}$. As superconductor films get thinner, $N \rightarrow 1$ and $\chi_{external} \rightarrow -\infty$, clearly an absurd result. In reality, as the film becomes thinner, and $N \sim 1$, flux immediately penetrates the film for any $H_{applied}$ and the superconductor is no longer in the shielding state. This yields broad transitions and a lack of a perfect diamagnetic state.

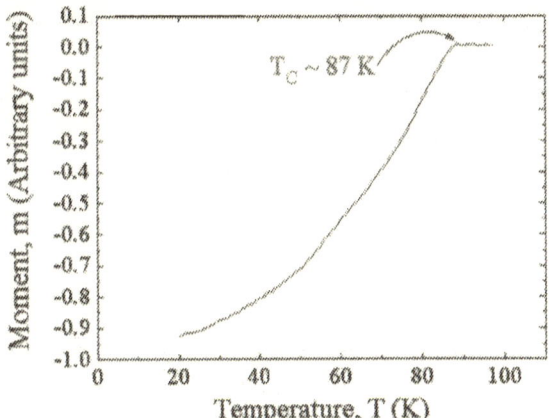

Figure 7.4 Moment versus temperature for a $Bi_2Sr_2CaCu_2O_{8+\delta}$ film obtained from SQUID magnetometry. Note the clear onset transition temperature of ~ 87 K.

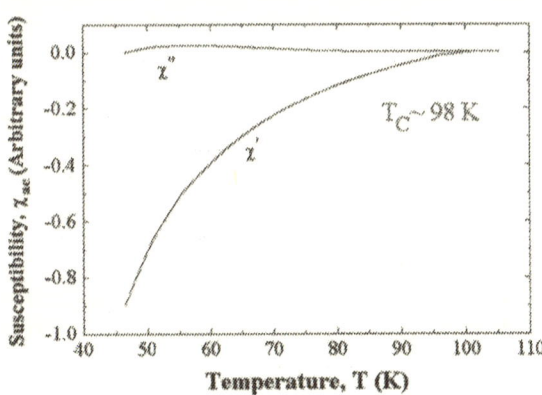

Figure 7.5 Real and imaginary parts of the AC susceptibility of an electrodeposited $Bi_2Sr_2CaCu_2O_{8+\delta}$ film versus temperature. The measurements were performed after a low temperature O_2 anneal. The superconductor onset transition occurs at ~ 98 K.

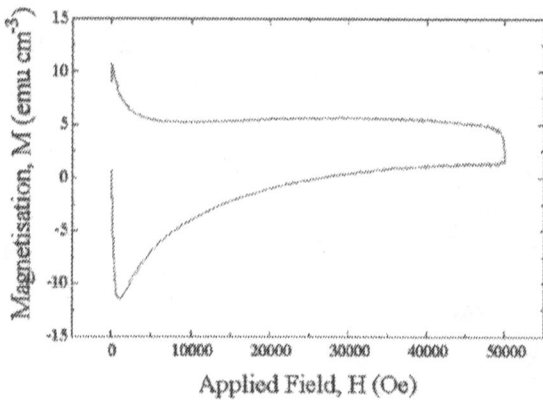

Figure 7.6 A typical M versus H loop for an electrodeposited $Bi_2Sr_2CaCu_2O_{8+\delta}$ superconducting film.

The $Bi_2Sr_2CaCu_2O_{8+\delta}$ films produced displayed good texturing, high phase purity, and with an excellent T_C. Measurements performed on a VSM were also used to determine the magnetic (intrinsic) critical current density, $J_{C,m}$, of the films. Initially, applied field sweeps were obtained to quantify the magnetic hysteresis of the films. Figure 7.6 is a typical sample magnetisation, M, versus applied magnetic field, H, loop for a bismuth-based superconducting film. The applied field was swept at a rate of 50 Oe s^{-1}, and the temperature maintained at 5 K during the sweeps. From the figure the hysteric effect is distinct. By applying the Bean model, also applied in Chapter 3 (and described in Appendix A.1), it is possible to quantify $J_{C,m}$, so long as the superconducting grain size is known. The magnetisation difference, ΔM, between the curve of increasing H and the curve of decreasing H is plotted, still against H. The Bean model is then applied, i.e. $J_{C,m} = 30\Delta M/D$, to obtain the magnetic critical current density of the films. The results for the $Bi_2Sr_2CaCu_2O_{8+\delta}$ films is shown in figure 7.7. It can be seen that a $J_{C,m}$ of 4.99×10^5 A cm^{-2} is obtained for an applied field of 1 T at a temperature of 5 K. Notice also that at an applied field of 4.5 T a $J_{C,m}$ of 2.19×10^5 A cm^{-2} is obtained indicating a promising field dependence on the magnetic critical current density.

Chapter 7: Superconducting Properties of Electrodeposited Films

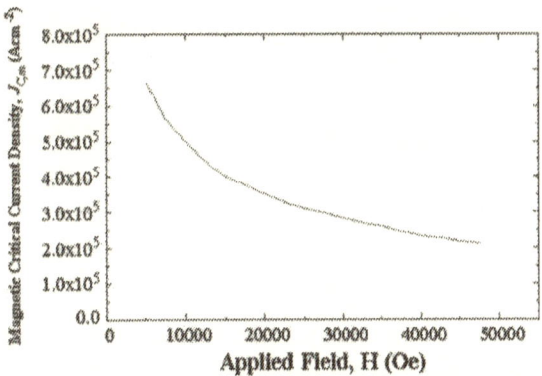

Figure 7.7 *The variation of magnetic critical current density, $J_{C,m}$, versus applied magnetic field, H, for an electrodeposited $Bi_2Sr_2CaCu_2O_{8+\delta}$ film.*

7.3 Superconductivity in Tl-Pb-Sr-Ca-Cu Films

The overriding difficulty in inducing superconductivity in the Tl-Pb-Sr-Ca-Cu, compared to the Bi-Sr-Ca-Cu films, is the loss of thallium during the sintering process. This problem occurs, to a lesser extent, with the lead also. To minimise material losses the films were sealed into quartz tubing in order to maintain the vapour pressures of the volatile materials at a level that would assist in film inclusion (see figure 7.8). The sample sits on an Al_2O_3 base about 5.0 mm. from a Tl-2223 pellet. The tube is closed with a piece of heat resistant wool, and then sealed, using an acetylene burner.

The samples were then sintered at a range of temperatures for 2 hours to form the appropriate superconducting phase. All samples were furnace cooled to room temperature at a rate of ~ 3.0 °C min^{-1}. The samples were then examined with EDS to determine the amount of thallium in the films. Figure 7.9 shows how the inclusion of thallium varied with furnace temperature. It should be noticed that the thallium content never achieved the ideal value of 0.5 (relative to Cu=3). At higher temperatures the film began to evaporate therefore limiting the amount of thallium that could be included in the films via this set up.

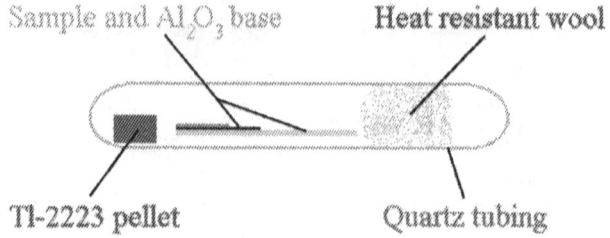

Figure 7.8 Schematic depicting the sample encapsulation, prepared for heat treatment.

The morphology of the heat treated films was very different from the $Bi_2Sr_2CaCu_2O_{8+\delta}$ films (figure 7. 10). The homogeneity was inferior when compared to the bismuth-based films, with what appeared to be a dense under layer covered randomly with small grains. The average feature size was ~ 5.0 µm, and again this was assumed to reflect the average size of the superconducting grains. The films were determined to be ~ 11.0 µm thick from SEM analysis. With this value the void percentage was re-calculated and found to be ~ 60 %, again indicating a reduction when compared to as-deposited films (~ 67 %). With the naked eye the films were matt black in colour and texture similar to the heat treated Bi-Sr-Ca-Cu films. Again the films were well adhered and ductile in nature.

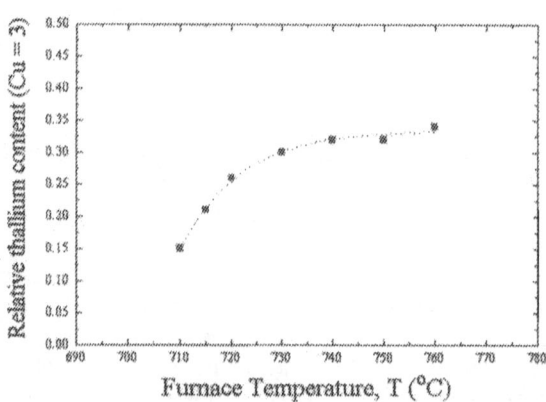

Figure 7.9 Thallium film content versus furnace temperature.

Chapter 7: Superconducting Properties of Electrodeposited Films

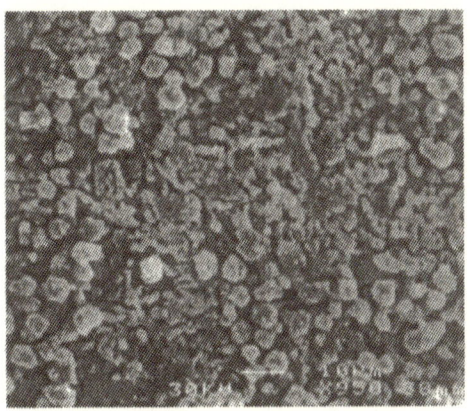

Figure 7.10 An SEM micrograph of a heat treated Tl-Pb-Sr-Ca-Cu film.

From XRD analysis it was found that the (Tl,Pb)-1212 phase formed first during heat treatment, followed by the formation of the desired (Tl,Pb)-1223 phase. Figure 7.11 shows the x-ray spectra predicted for polycrystalline (Tl,Pb)-1212, and (Tl,Pb)-1223 samples. These spectra were compared with the spectrum obtained for a reacted Tl-Pb-Sr-Ca-Cu film (figure 7.12), reacted at 770 °C. The lines in the heat treated film could mostly be indexed to the (Tl,Pb)- 1223 phase, but also to a lesser extent to the (Tl,Pb)-1212 phase, indicating that multi-phase films had been produced. Further optimisation of the heat treated procedure was unable to yield significant improvements to the phase development suggesting that an alternative procedure may provide the development required, possibly by way of using a two-zone furnace.

The texturing was not as high as that obtained for the bismuth-based films because of the presence of lines other then the 00l lines. The reduction in texturing may be attributed to the increased lattice mismatch between the Ag substrate and the film. Moreover, alignment may be limited because of the presence of the two phases.

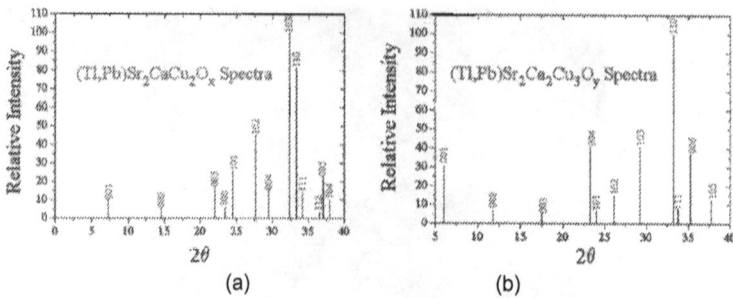

Figure 7.11 X-ray spectra of- (a) (Tl,Pb)-1212, and (b) (Tl,Pb)-1223.

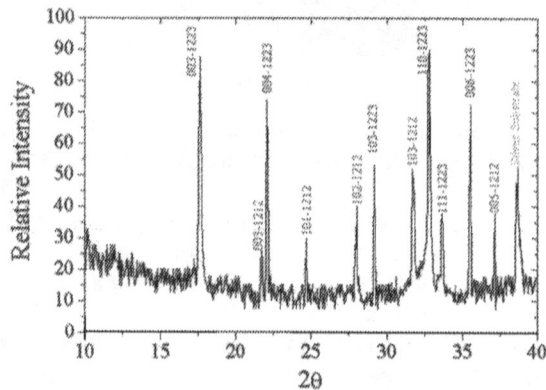

Figure 7.12 XRD spectra of a heat treated Tl-Pb-Sr-Ca-Cu film showing the existence of two different superconducting phases. The film was sintered at 770 °C for 2 hours in air and then furnace cooled to room temperature.

Further measurements to observe superconducting characteristics of the films were performed on a SQUID magnetometer and a VSM. SQUID measurements (figure 7.13) indicated a superconducting onset transition temperature of ~ 115 K consistent with the formation of a 1223 phase. No additional anneal was performed in an attempt to increase this value.

As with the $Bi_2Sr_2CaCu_2O_{8+\delta}$ films, measurements of the magnetisation of the films in an applied field were performed in order to obtain $J_{C,m}$ for the multi-phase thallium-based, films. It can be seen from figure 7.14 that the $J_{C,m}$ at 5 K in an applied magnetic field of 1 T was 1.26×10^6 A cm^{-2}. This value is noticeably smaller than the value obtained for the $(Tl_{0.5}Pb_{0.5})$-$(Sr_{1.8}Ba_{0.2})Ca_2Cu_3O_{9+\delta}$ powder produced in Chapter 3. It must be remembered, however, that the electrodeposited films were multi-phasic and therefore the value for $J_{C,m}$ is an average for the two phases synthesised. The field dependence of $J_{C,m}$ is encouraging with a value of 7.7×10^5 A cm^{-2} obtained in a field of 4.5 T and at a temperature of 5 K.

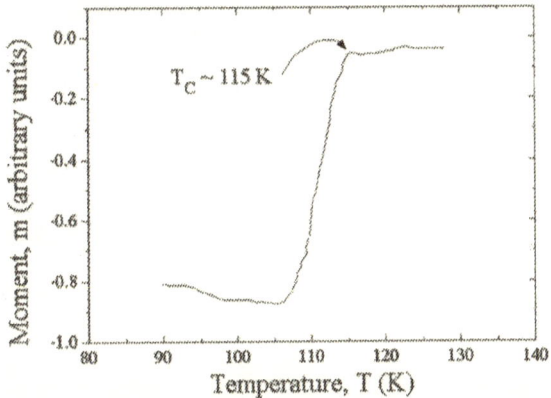

Figure 7.13 Moment versus temperature for a heat treated Tl-Pb-Sr-Ca-Cu film obtained from SQUID magnetometry. A $T_C \sim 115$ K is apparent.

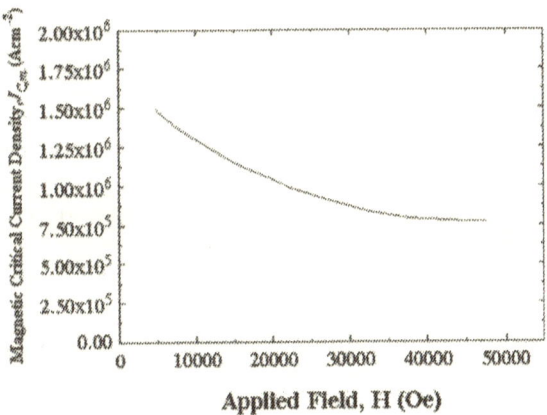

Figure 7.14 The magnetic critical current density versus applied field for a multi-phase electrodeposited Tl-Pb-Sr-Ca-Cu film.

7.4 Critical Current Densities in Electrodeposited Superconducting Films

Critical current measurements were performed on selected bismuth-based films. The measurements were performed using a technique similar to that described in section 2.8.

Contacts were made to the samples by applying a small amount of silver paint directly to the film. 'Blobs' of the conducting paint were added to provide two current leads and two voltage tapes. Whilst the paint was still wet wires were pushed into the paint and the sample was left to dry. When dry the sample, with wires, was placed into a tube furnace and cooked at 400 °C for 2 hours to ensure that the paint had diffused partially into the film to ensure a good contact.

The voltage resolution of the equipment was ~ 10 nV. The current vs.voltage response was observed to be linear at 77 K indicating that the current bad passed through the silver substrate and not the superconducting film. The resistance of the silver substrate was calculated to be 4.0×10^{-6} Ω at 77 K. An upper bound for the critical current density of the superconducting film may assumed to be 2.5 mA cm^{-2}, a disappointingly low value. Reasons for the unobserved $J_{C,t}$ may be that either the superconducting grains were poorly connected and/or the superconducting films were poorly adhered to the substrate.

7.5 Conclusions and Discussion

This chapter presented the superconducting properties of heat treated electrodeposited films of Bi-Sr-Ca-Cu and Tl-Pb-Sr-Ca-Cu. Appropriate sintering of the Bi-Sr-Ca-Cu films produced nearly single phase $Bi_2Sr_2CaCu_2O_{8+\delta}$ superconductors with an excellent level of texturing. The films also displayed good homogeneity over the entire surface with a typical grain size of 8.0 µm. The T_C values of these films was found to be 87 K which could be enhanced, via a low temperature oxygen anneal, to 98 K. These values are very promising indeed and represent a significant improvement over past attempts at producing $Bi_2Sr_2CaCu_2O_{8+\delta}$ films via an electrochemical stage. The bismuth-based superconducting films displayed a $J_{C,m}$ of 4.99×10^5 A cm^{-2} at 5 K in a field of 1 T. Furthermore, $J_{C,m}$ had an encouragingly low dependence upon the magnitude of the applied magnetic field, dropping by only 55 % at 4.5 T. Despite these encouraging intrinsic properties the transport properties of the films were disappointingly poor.

For the first time films of Tl-Pb-Sr-Ca-Cu have been manufactured. The heat treated films were multi-phasic composing of a mixture of $(Tl,Pb)Sr_2Ca_2Cu_3O_{8+\gamma}$ and $(Tl,Pb)Sr_2CaCu_2O_{8+\delta}$ with the 1223 phase being the majority phase. The films showed significant c-axis alignment and good homogeneity with an average grain size of ~ 5.0 µm. A T_C of 115 K was established, and a $J_{C,m}$ of 1.26×10^6 A cm^{-2} at 5 K in a field of 1 T. Again good field dependence of the magnetic critical current density was obtained with a decrease of only 40 % in an applied field of 4.5 T.

Both types of films have important roles to play in the development of superconducting devices. In particular, the Bi-2212 films could be used to form tapes for winding high field solenoids operating at 5 K, once a continuous process has been devised and proven. The attraction of the thallium-based films is the reduced anisotropy compared to other superconducting phases under consideration for applications. However, the weak link problem discussed in Chapter 3 would have to be addressed before successful competition against the bismuth-based candidates. In the initial investigations into the superconducting properties presented above limited transport measurements were made indicating low current carrying properties. In spite of this, the results have demonstrated that high quality material is indeed produced but further research is necessary to establish the current carrying capabilities.

The results presented in this chapter have formed the basis of two published papers [3,4].

References

1. M. Maxfield H. Eckhardt, Z. Iqbal, F. Reidinger, and R. H. Baughman, *Appl. Phys. Lett.*, 54 (1988) 1932.
2. S. J. Collocott, R. Driver, and C. Andrikidis, *Physica C*, 156 (1988) 292.
3. K. A. Richardson, D. W. M. Arrigan, P. A. J. de Groot, P. C. Lanchester, and P. N. Bartlett, *Electrochimica Acta*, 41 (1996) 1629.
4. K. A. Richardson, P. A. J. de Groot, P. C. Lanchester, and P. N. Bartlett, *Superlattices and Microstructures*, 21 (1997) 291.

Chapter 7: Superconducting Properties of Electrodeposited Films

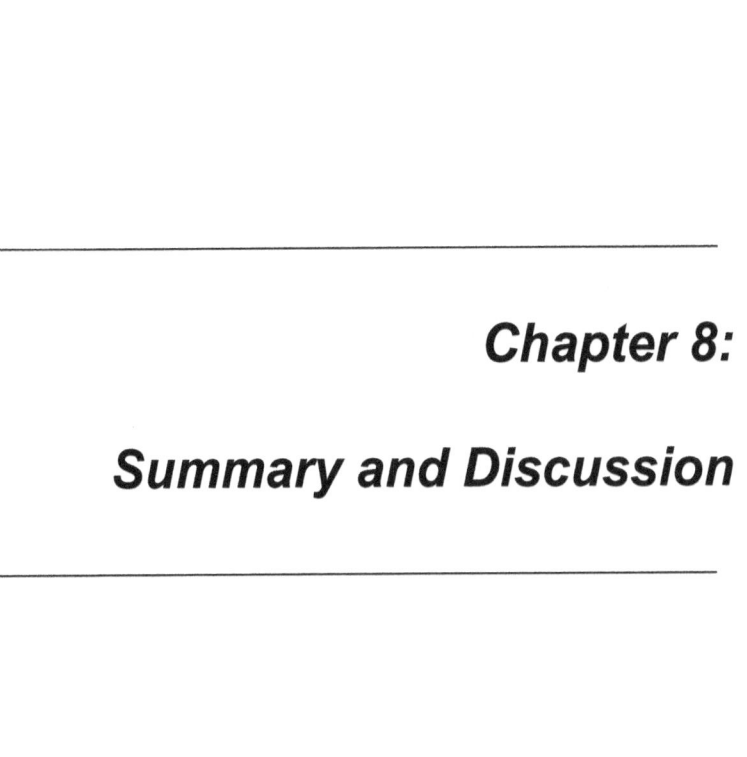

8 Summary and Discussion

8.1 Introduction

In this thesis the manufacture of long lengths of superconducting tapes and wires has been considered. In particular, fabrication of tapes via the PIT technique, using a newly discovered stoichiometry of the Tl-1223 superconducting compound was investigated, as well as an electrochemical approach. Promising results for both techniques were obtained and indications of future research directions uncovered.

8.2 PIT Fabricated Tapes

In Chapter 3 the fabrication of tapes based on the thallium-based superconductor $(Tl_{0.6}Pb_{0.2}Bi_{0.2})(Sr_{1.8}Ba_{0.2})Ca_2Cu_3O_{9+\delta}$ was presented. This particular powder was selected because of (i) the ease by which the Tl-1223 phase was obtained; (ii) the platelet-like morphology; (iii) the very high intrinsic current density, $J_{C,m}$, and; (iv) the good pinning properties of the material as observed through magnetic hysteresis measurements (see table 8.1). The research lead to a treatment process that yielded superconducting tapes with a promising $J_{C,t}$ and excellent microstructure.

Difficulties with thallium-based tape manufacture in the past have been linked to problems in obtaining phase pure powder, and the lack of texturing in PIT samples. The research reported herein has been demonstrated that single phase Tl-1223 powder can be manufactured reproducibly, and that texturing can indeed be found in the tapes produced. The texturing arises because of the favourable morphology of the as-synthesised powder. The research also hinted at the weak link problem inherent in the thallium-based compounds. This arose because, in spite of the excellent morphological properties of the tapes, transport critical current densities were comparatively low. The characteristics discovered, however, warrant further investigation and if this material is to be used to replace the successful bismuth-based tapes then the weak link issue will be an important area of investigation.

Chapter 8: Summary and Discussion

Tape Property	Result of research
Pinning	As-synthesised power displayed irreversible behaviour at applied fields below ~ 3 T.
Intrinsic critical current density	6.3×10^7 A cm^{-2} at 5 K, 1 T. 6.5×10^4 A cm^{-2} at 77 K, 1 T.
Superconducting transition temperature	~ 110 K.
Morphology	Well defined granular structure
Grain size	~ 18 µm.
Texturing	Partial c-axis alignment
Transport critical current density	Maximum of 5.6×10^3 A cm^{-2} at 77 K, 0 T.

Table 8.1 Characteristics of $(Tl_{0.6}Pb_{0.2}Bi_{0.2})(Sr_{1.8}Ba_{0.2})Ca_2Cu_3O_{9+\delta}$ tapes.

Companies like the American Superconductor Corporation, for example, have already produced working examples of superconducting magnetic energy storage (SMES) devices, rotating electrical machines, underground power transmission cables, transformers, current limiters, magnetic separation machines, magnets, and current leads. These devices were all built using the bismuth-based materials with engineering $J_{C,t}$'s of upto 4.4×10^4 A cm^{-2}. At present the bismuth-based superconductors provide sufficiently good qualities for rapid development of prototype demonstrations. Until researchers demonstrate that the thallium-based materials can surpass the bismuth-based materials in transport properties and ease of manufacture then companies will be reluctant to turn to the thallium-based materials with their associated toxic hazards. The research programme herein has shown that the thallium-based materials *can* be produced routinely, and safely with modest safety procedures. The promising characteristics of the tapes fabricated provide strong evidence that these materials have the potential of yielding superior conductors and therefore extending the range of operating parameters for future superconducting devices.

A proposed focus of future work in this area is described in detail in Chapter 3. The essence of future work should concentrate primarily on the problem of weak links between the grains and the optimisation of the reaction programme. Since the research herein was performed no significant progress in the area of

Chapter 8: Summary and Discussion

thallium-based PIT tapes has been reported indicating that a more concerted effort may be required in order to gain a competitive edge over already existing bismuth-based conductors.

8.3 Electrochemistry and Electrodeposition

A considerable amount of effort was expended in investigating both the basic electrochemistry of the relevant metals, and the preparation of superconductor precursor films via electrodeposition. In the electrochemical analysis, information was obtained concerning the reduction potentials, diffusion rates, and a variety of mechanisms that affect the behaviour of the elements that need to be combined to form the precursor films. The absence of literature meant that little understanding of the system employed was available. A summary of the results from the basic studies can be seen in table 8.2.

The details of the results are examined in detail in Chapter 4. The main outcome, however, is that the group of metals can be divided into two groups. The first group, containing Tl, Pb, Cu, Bi and Hg, are elements that are relatively easy to reduce, whereas the second group containing Sr, Ca and Ba, are difficult to reduce by comparison. Calcium in particular, a member of the second group, rapidly oxidises to form an insulating layer hindering further film growth, whereas the other elements form conducting layers. The significance of this is that thick films involving calcium cannot be grown without the presence of significant amounts of the other superconductor components (preferably copper) to ensure that the as-deposited film is conducting so that further deposition may occur. This causes problems because of the disparity in the respective reduction potentials. As explained earlier to produce dense, crystalline films, deposition must occur under kinetic rather than mass transport control. However, if co-deposition is being employed to deposit two or more elements then a potential must be selected to cause the simultaneous reduction of the electroactive species. This means that at least one of the species will be reduced with a potential very much larger than it's reduction potential, encouraging dendritic growth. A possible solution would be to deposit one layer from one bath, followed by another different layer from another bath. In principle this would work, but we must bear in mind that the aim is to develop a continuous process for the electrodeposition of superconductors. Having different baths would very much complicate the engineering involved in designing such a process, but more importantly it would yield low growth rates at high expense - two features that are highlighted as demonstrating favour toward an electrochemical route.

The method of co-deposition at one potential (constant-potential deposition) was investigated and found to yield film's with a void percentage as high as 90 %. Initial measurements on the film's superconducting properties were promising however, displaying excellent c-axis alignment, high phase purity, good magnetic critical current densities, and very high transition temperatures. However, if these films are to form conductors then a high void percentage is not ideal as it will yield low superconducting volume fractions.

Element	Electrochemical properties			
	E_{red} vs. Ag pseudo (V)	D_{CV} (10^{-6} cm^2 s^{-1})	D_{SV} (10^{-6} cm^2 s^{-1})	Type of process
Mercury (I)	~ 0.00	2.79	3.23	Quasi-reversible
Thallium (I)	-0.75	8.27	8.40	Quasi-reversible
Lead (II)	-0.55	1.05	2.48	Quasi-reversible
Bismuth (III)	-0.24	1.64	1.70	Quasi-reversible
Copper (I)	-0.38	3.63	7.97	Quasi-reversible
Barium (II)	-1.60	1.43	5.61	Irreversible
Strontium (II)	-1.80	-	-	Irreversible
Calcium (II)	-1.80	-	-	Irreversible

Table 8.2 Electrochemical Properties of the relevant elements

Pulsed-potential deposition was also investigated, and was found to yield a void percentage of ~ 67 %, significantly lower than constant-potential deposition. The pulsing of the potential increases film density by including a step that removes poorly adhered material causing a more even growth. During the -1.5 V to -1.0 V step a layer of the easily deposited metals is formed. Over this a layer of all the metals contained in the bath (say Tl, Pb, Sr, Ca, and Cu) is deposited at ~ -3.5 V. This high over potential causes the Tl, Pb, and Cu to be deposited at a high rate causing dendritic formation and hence high porosity films. When the potential is switched back again loose material is

Chapter 8: Summary and Discussion

removed to some extent. With this approach a dense film is produced followed by a porous film followed by a dense film, etc. Again, initial measurements performed on films produced this way were promising.

Deposition in the presence of an ultrasonic field is not a well researched area of electrochemistry. Ultrasound was found to yield the highest quality films though their stoichiometries were not comparable to the desired stoichiometry. The mechanical effects of the ultrasound removes loose material continuously throughout the deposition period. The increased velocity with which the cations impinge on the electrode causes denser film formation. The pulsed-potential approach was included so as to assist deposition by periodically laying down a relatively high conducting film. The void percentage of ~ 45 % is very promising but films were not reacted to obtain superconductors because of difficulties in obtaining the correct film stoichiometries. It was also found, through experience, that the solution should contain no more than 100 mM in total of the electroactive species. If this amount was exceeded then deposition occurred so rapidly most of the material did not adhere to the substrate. Extensive work is required to investigate the exact effects of an ultrasonic field in the deposition process.

The constant- and pulsed-potential techniques were used to fabricate a range of superconductor precursor films. Table 8.3 summarises the films produced, the deposition bath composition, and the technique applied. All deposition was performed at 30 °C, and in a dry, inert atmosphere. By adhering to the stringent experimental conditions discussed in Chapter 5 an excellent level of reproducibility was obtained. Several groups have published results concerning the electrodeposition of superconductors but few follow-up studies are reported. This may indicate the difficulties in achieving reproducible manufacture of the superconductors via this route. Once reproducibility has been obtained, however, it was not easy to maintain. The experimental procedure is extremely sensitive to environmental conditions, and neglecting any aspect of the procedure severely reduces the level of reproducibility. From the experience of the author, the most difficult aspect is the prevention of water contamination. Sometimes several days are required to prepare the dry box - a good indication being that

Chapter 8: Summary and Discussion

when P_2O_5 is placed in the box it remains powder for at least 36 hours. A better dry box would be required for further studies.

Films of Tl-Pb-Sr-Ca-Cu and Bi-Sr-Ca-Cu were heat treated to yield the correct superconducting phase. Tl-Pb-Sr-Ca-Cu films were found to be significantly c-axis aligned, with a multi-phase composition comprised from a majority of $(Tl,Pb)Sr_2Ca_2Cu_3O_{9+\delta}$ and a minority of $(Tl,Pb)Sr_2CaCu_2O_{7+\gamma}$. The mixed phase thallium-based films also displayed a T_C of 115 K and a magnetic critical current density, $J_{C,m}$, of 1.26×10^6 A cm^{-2} in zero field at 5 K. This is the first time that films of this type have been produced via an electrochemical route.

The Bi-Sr-Ca-Cu films displayed excellent c-axis alignment combined with very good phase purity. The Bi-2212 films had T_C's of upto 98 T - a world record for 2212 films fabricated this way - with $J_{C,m}$ of 4.99×10^5 A cm^{-2}. All the above results give credibility to the possibility of electrodeposition forming an integral part of all superconductor device manufacture. The ability to carry a significant current, though, was not demonstrated – a disappointing result.

Despite all the difficulties inherent in the electrochemical fabrication of superconducting films it has been shown that reproducibility can be achieved and good superconducting properties can result from the appropriate heat treatment programme. Further research is required if an electrodeposition approach is to out perform other more established techniques when considering qualities such as superconducting volume fraction and, more importantly, transport properties (per cross-sectional area) of the films produced. In fact, the large number of system parameters and the absence of high critical current densities may override the appeal to an electrochemical route as discussed above.

Chapter 8: Summary and Discussion

Precursor	Bath Composition	Technique	Film Stoichiometry
Bi-Sr-Ca-Cu	6.0 mM Bi(NO$_3$)$_3$ 5.0 mM Sr(NO$_3$)$_2$ 5.0 mM Ca(NO$_3$)$_2$H$_2$O 7.0 mM Cu(NO$_3$)$_2$H$_2$O	Constant -3.25 V 1800 s	2.0:1.8:1.2:2.0
Hg-Ba-Ca-Cu	4.2 mM Hg(NO$_3$)$_2$ 13.0 mM Ba(NO$_3$)$_2$ 11.5 mM Ca(NO$_3$)$_2$H$_2$O 3.0 mM Cu(NO$_3$)$_2$H$_2$O	Pulsed -4 V, -1 V 25 mins 18 °C	20.0:2.8:3.9:3.0
Ba-Ca-Cu	260.0 mM Ba(NO$_3$)$_2$ 130.0 mM Ca(NO$_3$)$_2$H$_2$O 46.0 mM Cu(NO$_3$)$_2$H$_2$O	Pulsed -3.25 V, -1 V 3600 s	1.7:2.1:3.0
Tl-Ba-Ca-Cu	5.6 mM TlNO$_3$ 500.0 mM Ba(NO$_3$)$_2$ 130.0 mM Ca(NO$_3$)$_2$H$_2$O 12.0 mM Cu(NO$_3$)$_2$H$_2$O	Pulsed -3.25 V, -1 V 3600 s	2.0:1.7:1.8:3.0
Tl-Pb-Sr-Ca-Cu	12.0 mM TlNO$_3$ 2.0 mM Pb(NO$_3$)$_2$ 200.0 mM Sr(NO$_3$)$_2$ 480.0 mM Ca(NO$_3$)$_2$H$_2$O 16.0 mM Cu(NO$_3$)$_2$H$_2$O	Constant -3.25 V 3600 s	4.8:0.6:1.9:1.8:3.0
Tl-Pb-Sr-Ca-Cu	12.0 mM TlNO$_3$ 2.0 mM Pb(NO$_3$)$_2$ 225.0 mM Sr(NO$_3$)$_2$ 450.0 mM Ca(NO$_3$)$_2$H$_2$O 16.0 mM Cu(NO$_3$)$_2$H$_2$O	Pulsed -3.25 V, -1 V 3600 s	4.6:0.5:1.6:2.1:3.0
Tl-Pb	8.0 mM TlNO$_3$ 1.5 mM Pb(NO$_3$)$_2$	Constant -1.3 V 1800 s	3.0:0.5
Tl-Pb-Cu	8.0 mM TlNO$_3$ 1.5 mM Pb(NO$_3$)$_2$ 8.0 mM Cu(NO$_3$)$_2$H$_2$O	Constant -1.3 V 1800 s	2.5:0.6:1.0
Sr-Ca-Cu	189.0 mM Sr(NO$_3$)$_2$ 330.0 mM Ca(NO$_3$)$_2$H$_2$O 24.3 mM Cu(NO$_3$)$_2$H$_2$O	Constant -3.25 V 3600 s	2.0:2.0:2.0

Table 8.3 *Electrodeposition parameters for different precursor films. All potentials are versus a Ag wire.*

8.4 Other Processing Methods

A variety of other production techniques have been employed in the production of superconducting films - in particular, wet chemical techniques. Wet chemical methods include dipping techniques, sol-gel processing and spray pyrolysis. The attractiveness of these methods arises not only from the ability to produce ceramics with high purity and good chemical homogeneity, but also because they are versatile for deposition coatings. Reasonable transport properties are also obtainable (see table 8.4).

Sol-gel and dip-coating are very similar techniques in that both techniques begin with a hot solution of, for example acetic acid, containing dissolved carbonates of the relevant elements [1]. The solution is continuously stirred to ensure a good level of mixing between the component materials. An additive is then introduced which increases the viscosity of the solution. A substrate is then dipped, or spun, into the solution resulting in a gel-like layer. The sample is then dried, leaving a solid, dark, amorphous, and possibly hygroscopic layer of constituent materials. In this, the precursor films are similar to those produced by electrodeposition, but denser. However, the homogeneity is simpler to control than when employing an electrochemical method.

Technique	Material	T_c (K)	$J_{c,t}$ (A cm^{-2})	Reference
Sol-gel	Tl-1234	122	3.5×10^2	1
Dip-coating	Bi-2212	-	2.7 A (I_c)	2
Spray pyrolysis	Tl-2212	102	5.0×10^3	3
Electrodeposition	Tl-1223	110	1.5×10^4	4

Table 8.4 A comparison of superconducting films characteristics for different preparation techniques. The transport properties are given at 77 K.

Once the precursor materials have been prepared, the samples are sintered under flowing oxygen to produce the superconducting phase. In the case of thallium-based superconducting films a thallium source is included to thallinate the sample. Transport critical current densities of long lengths of superconductor, produced in this manner, typically vary by 25-50 % using this approach. With carefully controlled heating, however, possibly utilising a multi-zone furnace, variations of the order of 10 % may be achieved [2].

Chemical spray pyrolysis is the simplest and least expensive method, beside vacuum and chemical vapour deposition. Moreover, it exhibits some advantages, such as high deposition rate, applicability for large and non-planar geometry coating, and easily controllable film composition - some aspects that are common with the electrochemical route. The films produced, though, suffer from comparatively low $T_{C,0}$ value, broad superconducting transitions, and low critical currents - other aspects that are associated with an electrodeposition stage.

The basis of spray pyrolysis is the creation of an aerosol generated from metallic salts. The aerosol droplets are very fine and produce a film of partially decomposed salts on the heated substrate (~ 500 °C). The surface is generally porous and heavily microcracked. This is mainly due to the fundamental difficulties in generating fine uniform droplets, and their thermal decomposition in terms of location, time, and gaseous environment. Efforts have succeeded though in producing crack-free films [5]. After precursor production the films are sintered in flowing oxygen in the presence of a thallium source for thallination, as with sol-gel, dip-coating and electrochemical techniques.

As mentioned previously, the ease by which these wet chemical techniques are controlled may prove to be the significant factor when deciding on what manufacturing technique is finally selected to produce long lengths of superconducting cable. The positive attributes of electrochemical production should inspire researchers to continue investigating the possibilities of implemented this technique for large-scale manufacture. There exists, though, difficult problems to solve before the future of electrochemical manufacture becomes clear.

8.5   Final Comment

The phenomena of superconductivity has enormous industrial potential, but the penetration of superconductor-based products has been slow thus far. An indisputable fact is that there are prizes, and profits, for the teams that solve the critical problems in understanding and processing these complex materials. The continuous processing of superconducting tapes and wires is a sought after goal that will open up the industrial market to the designers of a diverse range of time and energy saving devices.

The aim of this thesis was to investigate the fabrication process of the more established PIT technique, and the less well known electrochemical route. The PIT research yielded partially textured (a characteristic not found in previously reported thallium-based PIT tapes) and high purity tapes with promising transport properties. The in-depth study of manufacture using electrodeposition has produced a greater understanding of the processes involved and successfully demonstrated a reproducible procedure resulting in superconducting films. Extensive further work is required, but the impact upon society from superconducting devices draws ever nearer.

References

1. A. Wagner and G. Gritzner, *Supercond. Sci. & Tech.*, 7 (1994) 89.
2. J. Burgoyne, *Supercond. Bull.*, June (1996) 6.
3. M. Jergel, F. Hanic, G. Plesch, V. Štrbik, J. Liday, C. Falcony Guajardo, and G. S. Contreras Puente, *Supercond. Sci. & Tech.*, 7 (1994) 931.
4. R. N. Bhattacharya, P. A. Parilla, and R. D. Blaugher, *Physica C*, 211 (1993) 475.
5. S. P. S. Arya and H. E. Hintermann, *J. Less-Common Met.*, 164 (1990) 478.

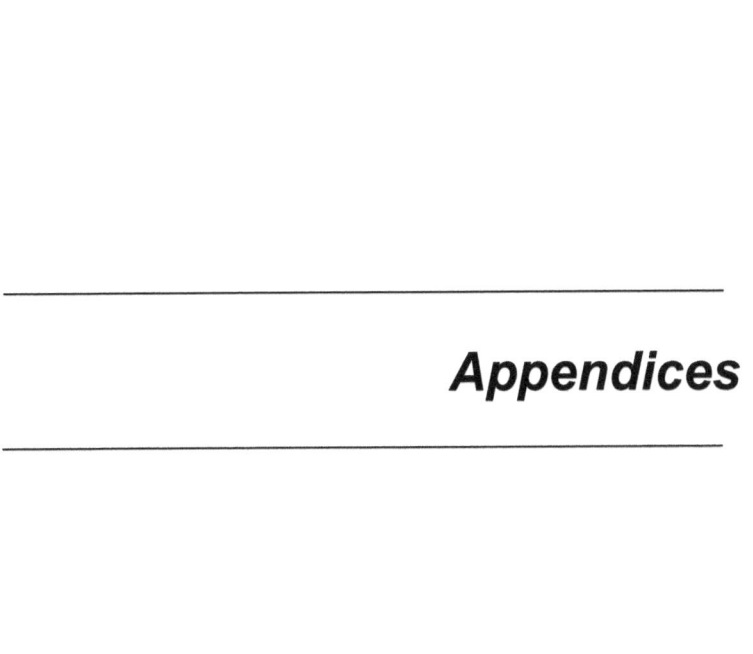

Appendix A – The Bean Model

Pinning in superconductors results in the irreversible behaviour seen in magnetic hysteresis experiments in which the response of the superconductor to a continuously swept field is measured. This hysteresis is linked with the critical current density by the Bean *critical state* model. When a magnetic field is applied to a superconductor, pinning in the sample prevents vortices, which enter at the surface, from immediately penetrating the whole of the sample. There is therefore a gradient in the flux line density, $dB/dx = \mu_0 j$. Bean argued that the shielding currents set up are always of equal magnitude to the critical current density, j_c, since for currents larger than j_c the Lorentz force will overcome the pinning force and cause the flux lines to move inwards. In the model it is also assumed that this current is independent of, or varies slowly with, the magnetic field. These assumptions result in a field profile (in this case for a slab sample) as shown in figure A1.

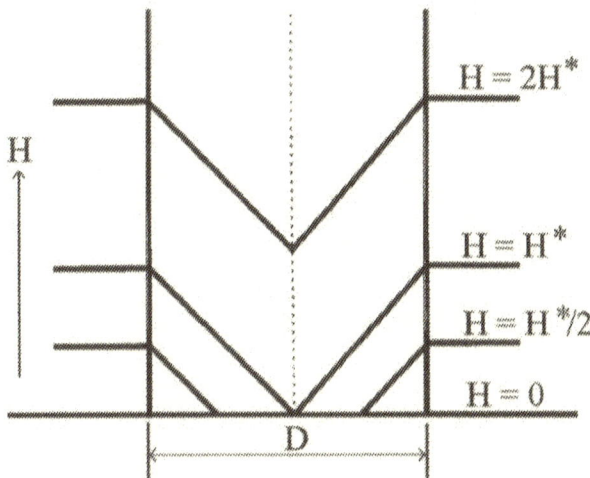

Figure A1 Field profile in a slab at various applied fields, as described by the Bean model.

Appendix A: The Bean Model

For $H_{c1} < H < H^*$ there is incomplete penetration of the sample. When $H = H^*$, the sample is fully penetrated and critical currents flow throughout the sample, and for fields greater than H^*, the field inside the sample simply increases.

Consider the magnetisation of a cylindrical sample as shown in figure A2, along with the field profiles for the points A, B, C, and D on the loop. At point A, the external field has just reached H^*. The field is further increased $H >> H^*$ along the path A-B, followed by a small decrease to point C. This decrease has lead to a local reversal of the field gradient and screening currents at the sample edge. The field is finally reduced to zero along C-D, but flux remains in the sample as screening current continuous to flow.

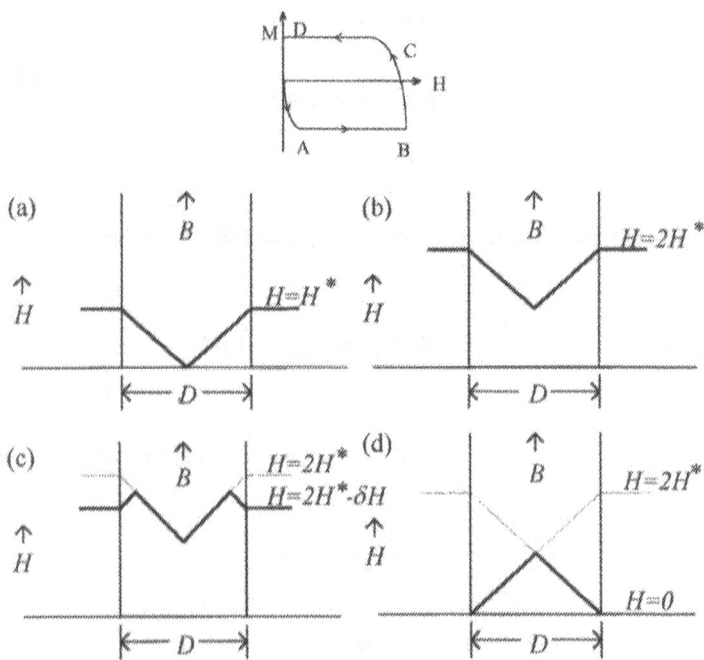

Figure A2 Magnetisation loop of a type-II superconductor, together with the field profiles at the points A, B, C, and D.

Appendix A: The Bean Model

Now the magnetisation, M, is given by,

$$M = \frac{1}{V}\iint_V \left(\frac{B}{\mu_0} - H\right) dV \qquad \{A.1\}$$

where B is the local field inside the sample. From the Maxwell equations we have,

$$\nabla \times \vec{B} = \mu_0 j_C \qquad \{A.2\}$$

This can be simplified for the case of a cylinder (assuming full penetration) if we take B directed along the symmetry axis.

$$\frac{\partial B}{\partial r} = \mu_0 j_C \Rightarrow B(r) = \mu_0 j_C r + b \qquad \{A.3\}$$

Since $B(R) = \mu_0 H$, where R is the radius of the cylinder, $b = \mu_0 H - \mu_0 j_C R$, and therefore,

$$B(r) = \mu_0 j_C (r - R) + \mu_0 H \qquad \{A.4\}$$

Using equation A.1, the measured magnetisation for an increasing field, $M_\uparrow$, is given by,

$$M_\uparrow = \frac{j_C}{\pi R^2} \int_0^R (r-R) 2\pi r\, dr = -\frac{j_C R}{3} \qquad \{A.5\}$$

Similarly for a decreasing field we obtain,

$$M_\downarrow = +\frac{j_C R}{3} \qquad \{A.6\}$$

The critical current density is therefore given by,

$$j_C = \frac{3\Delta M}{D} \qquad \{A.7\}$$

where $\Delta M = |M_\downarrow - M_\uparrow|$ and D is the sample diameter ($= 2R$). The units for ΔM and D are A cm^{-1} and cm, respectively. If ΔM is measured in emu cm^{-3} and D in cm then A.7 becomes:

$$j_C = \frac{30\Delta M}{D} \qquad \{A.8\}$$

Appendix B – Thickness of an Electrodeposited Film

The derivation related the total charge transferred during the deposition of a superconductor precursor films to the film thickness.

The total charge transferred during the deposition process is given by,

$$C_{total} = \int_0^{T_D} I(t)dt \qquad \{B.1\}$$

where T_D is the deposition time and $I(t)$ is the deposition current which is a function of time. We now invent a composite (imaginary) ion which is formed from all the different ions be co-deposited. The effective ion will have a charge, $Q_{effective}$, given by,

$$Q_{effective} = e \frac{\sum_{x=1}^{n} w_x q_x}{\sum_{x=1}^{n} w_x} \qquad \{B.2\}$$

where n is the number of different cations deposited, w_x is the atomic weight factor, and q_x is the cation charge in units of e, the electron charge. For example, if we consider a Bi:Sr:Ca:Cu = 2:2:1:2 film, the weighting factors are 2, 2, 1, 2, and the respective cation charges are 3 (Bi^{3+}), 2 (Sr^{2+}), 2 (Ca^{2+}), and 2 (Cu^{2+}). Hence, $Q_{effective}$ is ~ 2.29. The number of effective ions, $N_{effective}$, is defined as,

$$N_{effective} = \frac{\int_0^{T_D} I(t)dt}{Q_{effective}} \rightarrow \frac{\int_0^{T_D} I(t)dt \sum_{x=1}^{n} w_x}{e \sum_{x=1}^{n} w_x q_x} \qquad \{B.3\}$$

Appendix B: Thickness of an Electrodeposited Films

Next, we need to calculate the mass of one of the effective ions, $M_{effective}$. This is given by the expression,

$$M_{effective} = \frac{\sum_{x=1}^{n} w_x m_x}{\sum_{x=1}^{n} w_x} \qquad \{B.4\}$$

where m_x is the atomic mass number of the metals. For a Bi-2212 films:

$m_1 = [Bi] = 208.98$ g $m_2 = [Sr] = 87.62$ g
$m_3 = [Ca] = 40.08$ g $m_4 = [Cu] = 63.546$ g

and so $M_{effective} = 108.62$ g. The mass of the deposited material is therefore,

$$\frac{N_{effective} M_{effective}}{N_A} = \frac{\int_0^{T_D} I(t) dt \sum_{x=1}^{n} w_x m_x}{N_A e \sum_{x=1}^{n} w_x q_x} = \rho_{effective} V \qquad \{B.5\}$$

where N_A is Avogadro's Number, and $\rho_{effective}$ is the effective density of the deposited material and is given by the expression,

$$\rho_{effective} = \frac{\sum_{x=1}^{n} w_x \rho_x}{\sum_{x=1}^{n} w_x} \qquad \{B.6\}$$

For Bi-2212 the effective density ~ 6.29 g cm^{-3}. Hence, the volume, V, of the deposited film is,

$$V = \frac{\int_0^{T_D} I(t) dt \sum_{x=1}^{n} w_x m_x \sum_{x=1}^{n} w_x}{N_A e \sum_{x=1}^{n} w_x q_x \sum_{x=1}^{n} w_x \rho_x} \qquad \{B.7\}$$

If the area of the film is A_{film} then the thickness of, h_{film}, assuming that the film is evenly distributed over the substrate, is given by,

$$h_{film} = \frac{C_{total} M_{effective}}{A_{film} N_A e Q_{effective} \rho_{effective}} \quad \{B.8\}$$

Publications List

Papers

1. K. A. Richardson, S. Wu, D. Bracanovic, P. A. J. de Groot, M. K. Al-Mosawi, D. M. Ogborne, and M. T. Weller, "The Synthesis and Characterisation of $(Tl_{0.6}Pb_{0.2}Bi_{0.2})(Sr_{1.8}Ba_{0.2})Ca_2Cu_3O_{9+\delta}$ Powder and Ag-Sheathed Tape," *Supercond. Sci. Technol.*, 4 (1995) 238.
2. K. A. Richardson, D. W. M. Arrigan, P. A. J. de Groot, P. C. Lanchester, and P. N. Bartlett, "Electrodeposition of the Bismuth-Based Superconductor $Bi_2Sr_2CaCu_2O_{8+\delta}$," *Electrochimica Act*, 41 (1996) 1629.
3. K. A. Richardson, P. A. J. de Groot, P. C. Lanchester, and P. N. Bartlett, "The Electrodeposition of High Temperature Superconducting Films," *Superlatt. and Microstructures*, 21 (1997) 291.
4. K. A. Richardson, P. R. Birkin, P. N. Bartlett, P. A. J. de Groot, and P. C. Lanchester, "Towards the Electrochemical Manufacture of Superconductor Precursor Films in the Presence of an Ultrasonic Field," *J. Electroanal. Chem.*, 420 (1997) 21.

Conference Presentations

1. K. A. Richardson, D. W. M. Arrigan, M. Al-Mosawi, D. M. Ogborne, P. A. J. de Groot, P. C. Lanchester, C. Beduz, P. N. Bartlett, and M. T. Weller, "The Manufacture of Thallium-Based Superconducting Wires, Tapes, and Films," *Physique en Herbe 94*, Montpellier, July 4-8, 1994.
2. K. A. Richardson, S. Wu, D. Bracanovic, P. A. J. de Groot, D. M. Ogborne, and M. T. Weller, "A Study of $(Tl_{0.6}Pb_{0.2}Bi_{0.2})(Sr_{1.8}Ba_{0.2})Ca_2Cu_3O_{9+\delta}$ Powder and Ag-Sheathed Tape," *High Temperature Superconductors VIII*, Birmingham, 19-21 September, 1994. Also presented at *Condensed Matter and Materials Physics 94*, Warwick, 19-21 December, 1994.
3. K. A. Richardson, D. W. M. Arrigan, P. A. J. de Groot, P. N. Bartlett, and P. C. Lanchester, "Morphology and Reproducibility of Electrodeposited $(Tl,Pb)Sr_2Ca_2Cu_3O_{9+\delta}$

Thick Films," *Condensed Matter and Materials Physics 94*, Warwick, 19-21 December, 1994.

4. K. A. Richardson, D. W. M. Arrigan, P. N. Bartlett, and P. A. J. de Groot, "Electrodeposition of Thick Films of Superconductor," *Electrochemistry 95*, Bangor, September, 1995.

5. K. A. Richardson, P. A. J. de Groot, P. C. Lanchester, P. N. Bartlett, "The Electrochemical Synthesis of High Temperature Superconducting Films," *5th World Congress on Superconductivity*, Budapest, 7-11 July, 1996.

www.ingramcontent.com/pod-product-compliance
Lightning Source LLC
Chambersburg PA
CBHW030919180526
45163CB00002B/400